TECTONIC
PROCESSES

PROCESSES IN PHYSICAL GEOGRAPHY

Editor: Darrell Weyman

TECTONIC PROCESSES

Darrell Weyman

Blandford Upper School
Blandford
Dorset

London
GEORGE ALLEN & UNWIN
Boston   Sydney

First published 1981

GEORGE ALLEN & UNWIN LTD
40 Museum Street, London WC1A 1LU

British Library Cataloguing in Publication Data

Weyman, Darrell Richard
 Tectonic processes. – (Processes in physical
geography; no. 4).
 1. Geology, Structural
 I. Title II. Series
 551.1′36   QE601   80-41834

 ISBN 0-04-551044-X

Set in 10 on 11 point Times by Servis Filmsetting Ltd, Manchester
and printed and bound in Great Britain by
William Clowes (Beccles) Limited, Beccles and London

Preface

The study of the distribution of the major features of the Earth's surface was traditionally a major part of any school geography course. The examination of these features tended, however, to be highly descriptive and frequently lacked much in the way of real intellectual stimulation. The study of volcanoes and earthquakes, whilst fascinating in themselves, also occupied an awkward place within the geography syllabus with the feeling commonly expressed that this was, perhaps, more properly of interest to the geologist. It is hardly surprising, therefore, that the advent of quantitative methods with a consequent improvement in explanatory theory in other areas of physical geography such as geomorphology has led to a general loss of interest in global physiography. This, in my opinion, is a situation much to be regretted. It is also my firm belief that the development by geologists of plate tectonics (or global tectonics if you prefer) as a comprehensive theory has provided the geographer with a clear, coherent and totally absorbing framework within which to view world physiography. The aim of this book is, therefore, to describe and *explain* the origin and distribution of the major features of the Earth's surface by making use of our current understanding of tectonic processes. In order to achieve that aim, it is necessary to present the evidence found at and beneath the Earth's surface which has been used to investigate those processes. As such, this might be described as a geology book, but I make no apologies for leaning heavily upon the terminology and concepts developed by geologists, since I believe the geographer has a duty to understand as many aspects of this planet upon which we live as may be possible.

DARRELL WEYMAN
Fiddleford

Acknowledgements

In writing a book of this type one inevitably leans very heavily upon an enormous range of source materials for both text and illustrations. It is probably impossible to fully acknowledge the debt which I owe to so many authors. In general terms, however, I would wish to mention the books listed under 'Further reading' at the end of this volume. More particularly, I would like to thank the following for permission to use original diagrams as the basis of some of my illustrations:

J. R. L. Allen, Figs 1.4, 4.15 and 7.5 from *Physical geology* (George Allen & Unwin 1975);

D. L. Anderson, for maps of the San Andreas fault from 'The San Andreas Fault', *Scientific American* (1971);

R. S. Dietz and J. C. Holden, maps of continental breakup from 'The breakup of Pangea', *Scientific American* (1970);

J. R. Heirtzler and W. B. Bryan, maps of the Atlantic seafloor from 'The floor of the Mid-Atlantic Ridge', *Scientific American* (1975);

R. J. Rice, Figs 5.2 and 5.5 from *Fundamentals of geomorphology* (Longman 1977);

D. H. and M. P. Tarling, Figs 13, 16, 18, 22 and 37 of *Continental drift* (Bell 1971).

I have also drawn material for diagrammatic purposes from the following sources:

E. Bullard, 'Seafloor spreading', *Scientific American* (1969);

F. M. Bullard, *Volcanoes: in history, in theory, in eruption* (University of Texas Press 1962);

K. C. Burke and J. T. Wilson, 'Hot spots on the Earth's surface', *Scientific American* (1976);

S. P. Clark, *Structure of the Earth* (Prentice-Hall 1971);

J. F. Dewey and J. M. Bird, 'Mountain belts and the new global tectonics', *Journal of Geophysical Research* (1970);

J. J. Donner, 'Land/sea level changes in Scotland', quoted in A. S. Goudie, *Environmental change* (Oxford University Press 1977);

R. W. Fairbridge, 'Eustatic changes in sea level', in *Physics and chemistry of the Earth* (Van Nostrand Reinhold 1961);

P. Francis, *Volcanoes* (Penguin 1976);

Heirtzler, Le Pichon and Baron (1966), quoted in S. P. Clark, *Structure of the Earth* (Prentice-Hall 1971);

A. Holmes, *Principles of physical geology* (Nelson 1965) (material from R. Staub);

B. Isacks, J. Oliver and L. R. Sykes, 'Seismology and the new global tectonics', *Journal of Geophysical Research* (1968);

M. N. Toksoz, 'The subduction of the lithosphere', *Scientific American* (1975).

I should like to take this opportunity to apologize in advance for any inadvertent use of original material without permission.

I am deeply indebted to the comments made on the first draft of this book by Nigel Bates, Bruce Marsh and Chris Wilson. I have incorporated many of their suggestions in the final version, although the shortcomings and inadequacies that doubtless still exist are of my own making. As ever, I am also greatly indebted to the support and guidance given by Roger Jones, without whom this book would never have appeared.

R. D. W.

Contents

List of Tables

Chapter 1

Introduction

1.1 Major features of the surface of the Earth

A casual glance at a world map in almost any atlas will be sufficient to make it clear that the geographer regards sea level as a rather fundamental marker point. Those areas of the Earth's surface below sea level are virtually all covered by water and are generally coloured blue on a map, whereas 'dry land' appears in various shades of green and brown when the relief is to be shown. Even when the surface relief below sea level is shown, the shading is still in tones of blue, and it is not really possible to confuse areas of land and sea. If you asked a geologist to draw a world map, however, the result might well look a little more like Figure 1.1. Although this map may initially appear familiar, the overall colour effect is designed to divide the surface into **continental** and **oceanic** areas. The continental areas defined in this manner are slightly larger than the land areas of the geographer's map, because they include the shallow-water **continental shelves**, which lie marginal to the land masses in some places.

The geologist makes this fundamental distinction between continental and oceanic because, as we shall see later, the age and type of rocks underlying the two areas are somewhat different. The distinction also has geographical significance, however, and this can be seen most clearly by examining the relief of the Earth's surface. Figure 1.2a is a histogram showing the total surface area of the Earth lying at different levels above and below sea level. As may be expected, only a small part of the Earth's surface lies at the two extremes of altitude. On the one hand, **mountain chains** rise above sea level to heights of almost $+10$ km, whereas below sea level the deep **ocean trenches** descend to below -10 km. Together, however, these amount to no more than about 10 per cent of the total surface area, leaving most of the surface lying between $+2$ km and -6 km. As Figure 1.2a makes clear, however, this huge area falls into two distinct levels:

a higher one lying between $+2$ km and -1 km, and a lower level between -3 km and -6 km. This pattern is perhaps seen even more clearly if Figure 1.2a is redrawn as a cumulative frequency curve of altitude (Fig. 1.2b), which is commonly known as a **hypsographic curve**. The higher level obviously represents most of the land areas of the world plus the continental shelves, and the lower level represents the floor of the oceans. This altitude grouping provides a purely physical reason for distinguishing continental and oceanic areas. The small surface area between these two levels, shown by the steep central portion of Figure 1.2b, comprises the relatively steep slope connecting the edge of the continental shelves with the ocean floor and also the shallower areas of the oceans known as the **ocean rises**. It must be emphasised that Figure 1.2b in no way represents the geographical distribution of altitude. To make this point clearer, Figure 1.3 is a very generalised cross-section from the Pacific Ocean to Africa, on which the various altitude categories are shown in something like their true geographical setting.

The major features of the Earth's surface are not only grouped by altitude; they are also grouped in space. Reference to Figure 1.1 should remind the reader that the continental areas are themselves rather obviously grouped: first, within the northern hemisphere; and also into the two blocks comprising the 'Old World' of Europe, Asia and Africa and the 'New World' of the Americas. As we have already seen, the surface of the continents lies largely below 2000 m, and much of this is either flat or has only low relief. This includes landscapes such as the rolling prairies of North America or the steppes of Asia, the high tablelands of Africa, the flat coastal plains of the Gulf of Mexico and the offshore continental-shelf areas of North-West Europe or South-East Asia. In many places hills and mountains stand above the surrounding lowlands, but even areas such as the Appalachians of North America or the Urals of central Russia are

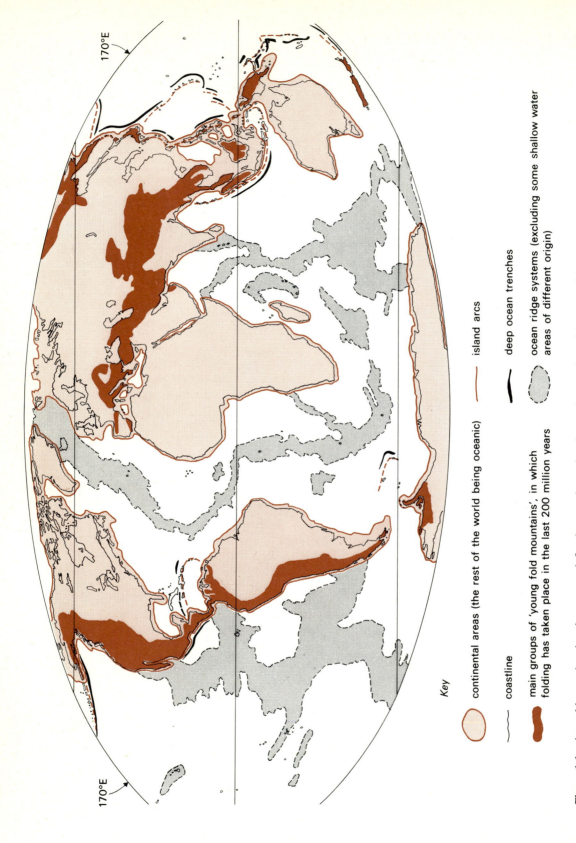

170°E

170°E

Key

⬭ continental areas (the rest of the world being oceanic)

— coastline

▬ main groups of 'young fold mountains', in which folding has taken place in the last 200 million years

— island arcs

— deep ocean trenches

⬭ ocean ridge systems (excluding some shallow water areas of different origin)

Figure 1.1 A world map showing those areas defined as continental and oceanic, together with some of the major linear structures of the Earth's surface.

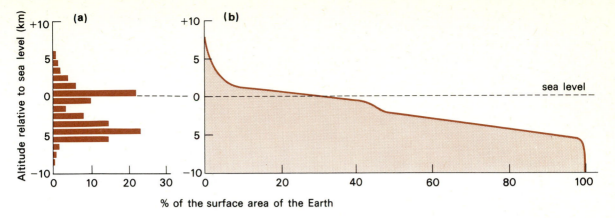

Figure 1.2   World relief: (a) A histogram showing the proportion of the world's surface at various altitudes relative to sea level. (b) A cumulative frequency curve of world altitude, known as the hypsographic curve.

really no more than the eroded relics of former mountains when compared with the major mountain chains of the Earth. The mountain chains shown in Figure 1.1 are commonly referred to as **young fold mountains**, since they have been formed in the recent geological past, even if they include much older elements. These mountain belts include virtually all of the surface area above 2 km, and it is noticeable that they do not occur in random groups over the Earth but form long continuous chains. The two most important are: (a) the Andes–Rocky Mountain system, forming a line down the western seaboard of the Americas; and (b) the Alpine–Himalayan system, which forms a more or less continuous west–east chain between Europe and

Asia to the north and Africa and India to the south. Although mountains form the most prominent linear feature on the continental surfaces, some continents are cut by linear depressions in the form of very long, deep and steep-sided valleys. Since the sides of these valleys are formed by the faulting, or 'rifting', of surface rocks, they are known as **rift valleys**. By far the most famous rift-valley system is that found in East Africa, which stretches from the Ethiopian Highlands in the north to Lake Nyasa in the south.

Where mountain chains run parallel to the coastline, as in the case of the Andes, offshore the seabed slopes steeply to the ocean floor at about -3 km, and there is no real continental shelf. In this case

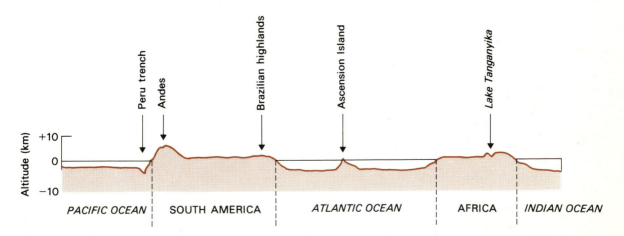

Figure 1.3   A generalised section from Africa through South America to the Pacific, showing the geographical relationships of major surface features.

the edge of the continental unit as a whole is close to the coastline (Fig. 1.1). Where a coastal plain meets the sea, however, it is often continued offshore as a wide continental shelf before the continental slope, and therefore the edge of the continent proper, is met. The 'edge' of Europe, for example, is found some 400 km west of Ireland. Geologically speaking, as we have already intimated, continental shelf and coastal plain are more or less identical. They are part of the single structural unit that we are calling a continent, across the edge of which sea level migrates backwards and forwards over time as the volume of water in the oceans changes with the growth and decay of ice sheets. Although the shoreline seems to be so permanent a marker in human terms, over geological time its position is constantly changing.

The oceanic areas themselves show just as much relief as the continents. In places the ocean floor between −3 km and −6 km is a true plain, but frequently the bottom has considerable relief. The **abyssal plains** (i.e. the area between −3 km and −6 km) are frequently interrupted by rises and depressions beyond this altitude range. Some rises are irregular groupings that form **sea mounts**, or islands if they reach sea level. More significant, however, are extensive linear systems on a scale analogous to that of the continental mountain chains. The first of these is the so-called **ocean ridge** (Fig. 1.1). These ridge systems are virtually submarine mountain chains, several hundred kilometres wide, which rise to a ridge usually less than −2 km below sea level and frequently form islands. The most famous of these systems runs down the centre of the Atlantic, its course paralleling the adjacent coastlines perfectly, breaking the surface of the ocean at Iceland, the Azores, Ascension Island and other points. Similar features can be found in the Indian and Pacific Oceans, but in these cases the ridge is nothing like central to the ocean basin. The other linear feature of the oceanic areas is the ocean trench, which has already been identified in the discussion on surface altitude. Reference to Figure 1.1 will show that the trenches are long, relatively narrow features descending to below −10 km. They are grouped in linear fashion. The trench lying off the western coast of the Americas clearly runs parallel to the Andes–Rocky Mountains chain, and most other trenches in the world seem to lie parallel to groups of volcanic islands, which so often take on an arcuate plan that they have become known as **island arcs**. The clearest trench–island arc systems are found in the western Pacific in groups such as the Aleutian Islands, the Kuril Islands and the Marianas Islands.

1.2   The purpose and organisation of this book

This brief summary of the major features of the Earth's surface should serve to indicate two things to the reader: first, that the surface of the Earth is very diverse, and second, that there is a pattern in the way in which the features are distributed around the world. It is the purpose of this book to describe these features in more detail and to offer a reasonable explanation for their origin and distribution. For some reason of historical accident, geographers seem quite happy to discuss the processes by which landscapes develop under the operation of agents of erosion (the province of geomorphology) but feel that an understanding of **tectonic processes** (i.e. those having their origin beneath the Earth's surface) remains the exclusive right of the geologist. This is a viewpoint with which I have little sympathy. If to understand the Earth's surface means using the geologist's knowledge, then so be it.

In recent years, geologists have developed the theory of plate tectonics which is an attempt to explain many of the Earth's major surface features through an understanding of the processes which operate deep beneath the surface. In this book we shall explore tectonic theory, by presenting the evidence upon which the theory has been developed. We shall start by looking at the evidence for continental drift which proved to be of crucial importance in the development of the more general theory. This is followed by a discussion of the evidence which has improved our understanding of the nature of the Earth's interior and, thereby, has suggested a possible mechanism to explain the movement of the continents. We shall then return to the substance of the book by describing and attempting to explain the surface features found in different **tectonic environments** around the world.

During this journey of exploration we shall inevitably have to employ some geological terminology. As far as possible, technical words will be explained as we go along, but it is necessary to assume that you, the reader, have at least a passing acquaintance with the words used to describe rocks, minerals and geological time. If you have not met words such as 'igneous' or 'sedimentary', 'granite' or 'sandstone' before, it might be as well to glance at the Appendices provided at the end of the book now. If, however, you have at least the barest knowledge of basic geology then read on.

Chapter 2

Continental Drift

2.1 Introduction

At first, the notion that the continents are not fixed upon the surface of the Earth but wander across the globe seems preposterous. Nonetheless, the idea has periodically been proposed over the course of several centuries by scientists who have felt that the geological evidence can be interpreted in no other way. In 1912 Alfred Wegener, a German scientist, first collected together evidence from a number of different sources and thus earned himself the title of the originator of modern drift theory. Even so, Wegener's ideas were met with scepticism or outright disbelief, and it was some time before the continuing accumulation of data became increasingly difficult to ignore. In this chapter we shall examine the nature of that evidence and begin to draw some conclusions about the movement of the continents. We shall, for the moment, ignore some of the recent seafloor evidence and keep strictly to what may be observed on the continents themselves.

2.2 The evidence for continental drift

2.2.1 Shape of the continents

It was in 1620 that Francis Bacon first observed that the eastern and western margins of the Atlantic Ocean seem to fit together in a way that can scarcely be accidental. Since that date many other writers have made the same observation, but it is by no means easy to make a convincing map to show that the Americas fit exactly against the coastline of Europe and Africa. In the case of Figure 1.1, for example, any attempt to place the shorelines together would result in numerous gaps or overlaps. There are two main reasons for this difficulty:

(a) Any map is an attempt to represent a curved surface on a flat page, and ultimately something has to be distorted to achieve that end. In some map projections the distances are not on a fixed scale, and in others it is the shape that suffers; but in either case a compromise is reached, which does not help with the problem of continental shape-fitting. Needless to say, the same difficulties attend all the maps in this book, even where care has been taken to minimise the problem.

(b) The coastlines of the continents may have suffered considerable alteration since splitting took place, as a result of erosion, deposition, volcanic activity or mountain-building. In the case of the Atlantic, for example, areas of river delta (e.g. the mouth of the Niger) have been built up by deposition in the recent geological past.

Within the last few years, various attempts have been made to improve the techniques of continental jigsaw-puzzle work. In some cases globes with movable surface features have been built; in others computers have been employed to fit continents in such a way as to leave a minimum of either gap or overlap. These 'optimisation' methods have been used most successfully by Sir Edward Bullard, who found, significantly, that the best Atlantic fit can be achieved not by using the coastline but by using the 2000 m below sea level contour (Fig. 2.1, but see also Fig. 4.1 for present-day Atlantic contours). The important point about this discovery is, of course, that the fitting is achieved using the edge of the continental areas defined in Section 1.1 by altitude frequency rather than using the shoreline. There can be little doubt that the maps produced by Bullard (e.g. Fig. 2.1) are fairly convincing, and it is difficult to imagine how this shape similarity could be produced by random variations. It really does look as though the two sides of the Atlantic were once joined. Equally interesting is the prospect of carrying out shape-fitting exercises on other continents to produce maps like that shown in Figure 2.6a.

2.2.2 Structural similarities

In some cases continents, now separated by an ocean, appear to possess areas that are remarkably similar in terms of the age, origin and folding of the rocks that underlie them. A good part of the southwestern Sahara (Mali and the adjoining coastal states), for example, is underlain by rocks that are identical to those found in the Guyana highlands of South America. In the North Atlantic the situation is even clearer. Much of the upland areas of Britain was formed during two episodes of mountain-building (Section 6.5). The first, the Caledonian (Appendix B), primarily affected deep water sediments and volcanic rocks of Lower Palaeozoic age (570–395 million years ago), which were folded into mountains along a south–west to north–east axis through central and northern Wales, the Lake District and Scotland. The second mountain-building period, the Hercynian, caused shallow-water continental-shelf deposits of Upper Palaeozoic age (395–225 million years ago) to be

folded gently to form the Pennines and southern Wales coalfields. Further south in Devon and Cornwall, deeper water deposits of the same age were intensely folded, and granite masses (e.g. Dartmoor) were intruded into the rocks. Both these mountain systems can be traced beyond Britain. Remnants of the Caledonian mountains can be found in Norway, eastern Greenland, Newfoundland, New England and the western Sahara. If the structural lines (i.e. the fold axes) of this single mountain system are drawn upon a reconstructed map of the North Atlantic seaboards (Fig. 2.2), the essential unity of the mountain chain becomes clear. A similar pattern emerges from the line of granite masses associated with the Hercynian mountain chain. Granite masses of comparable age can be found along a belt stretching from the Sudeten Mountains of Central Europe, through southwestern England to the Appalachian Mountains of eastern North America.

2.2.3 Reconstruction of past climatic patterns

The reader will doubtless be aware that the climate of this planet is in a state of constant change and that in the past there were periods when the average temperature of the Earth was both higher and lower than at present. It is therefore not particularly surprising that the rocks beneath the surface often provide evidence of climatic change. Sediments can often be used for this purpose, but the fauna and flora found in fossil form in sedimentary rocks provide a more obvious clue to the climate of the past.

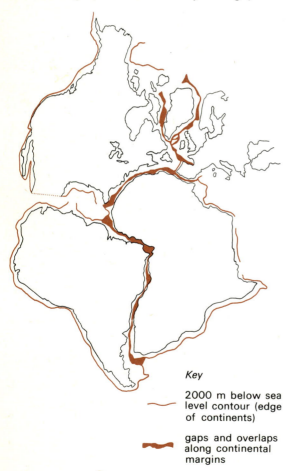

Key

— 2000 m below sea level contour (edge of continents)

— gaps and overlaps along continental margins

Figure 2.1 A reconstruction of the fit of continents on either side of the Atlantic, based upon the outline at 2000 m below sea level (after Bullard 1969).

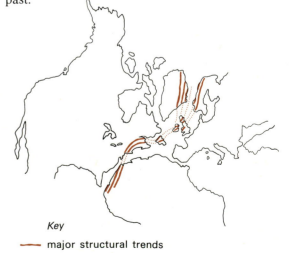

Key

— major structural trends

Figure 2.2 The main structural lines of the Caledonian mountains, shown on a reconstruction of the North Atlantic (from Tarling & Tarling 1972).

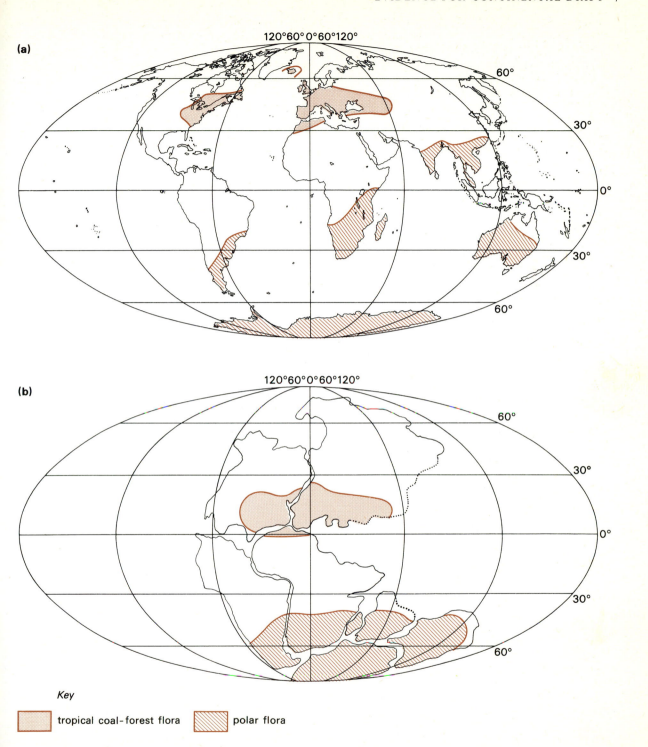

Figure 2.3 (a) The distribution of tropical coal forest and polar fossil floras in rocks of Carboniferous age. (b) A reconstruction of world geography in Carboniferous times to give a climatic pattern similar to that of the present day (from Tarling and Tarling 1972).

In the present context, a rather interesting picture emerges if the global pattern of climate is reconstructed for one period in the past. Figure 2.3a, for example, shows the world distribution of rocks of Carboniferous age (about 300 million years ago), containing either a tropical swamp flora or a polar flora. At first sight it seems possible that a shift in the climatic belts may have taken place, which can possibly be explained by a variation in the position of the Earth's axis. On closer examination, however, it seems that during the Carboniferous period almost half of the globe was experiencing Arctic conditions, while a substantial section of the northern hemisphere remained tropical. Comparison with a modern map of world climates should make it clear that it is not easy to imagine what happened to the northern polar area nor how it was possible for polar and tropical climates to be so close to each other. In any case, such a large polar area suggests an overall cooling of the Earth, which makes the presence of the tropical area difficult to explain. Faced with these difficulties, it seems preferable to move the continents around until the Carboniferous climates approximate the present day distribution (Fig. 2.3b). To base a reconstruction purely on the distribution of palaeoclimates is obviously asking for trouble, but Figure 2.3b is, of course, compatible with both the shape and the structural evidence already discussed.

2.2.4 The pattern of evolution

The evolution of botanical and zoological species into ever more complex forms is the basis of the theory normally attributed to Charles Darwin. As far as marine organisms are concerned, evolutionary changes are usually transmitted rapidly throughout the planet, although there are instances of polar forms being unable to cross the warmer tropics and vice versa. Many land organisms, however, especially reptiles, mammals and certain plants, have the greatest difficulty in crossing any distance of water, with the result that they will spread rapidly only across continents connected by land bridges. If a continent is separated from all the others, its fauna and flora may develop in a quite distinctive manner, as was demonstrated most graphically by the discovery of Australasia.

As far as the present discussion is concerned, the evolution of one group of organisms may throw considerable light upon the pattern of continental drift. Consider, for example, the case of the mammals. Mammals may be divided into the placental group and the marsupial group. Placental mammals include man and other animals where the offspring are independent of the mother from birth, whereas the marsupials (e.g. kangaroo) retain their offspring in a pouch outside the mother's body after birth. Generally, marsupials are herbivores with a fairly placid nature, in contrast to the often more aggressive placental mammals, many of which are, of course, carnivores. The available evidence seems to suggest that, where placentals and marsupials are in competition, the placentals always prove more successful and dominate an environment, leading to the elimination, or at least severe reduction, of the marsupial population.

From the fossil record it appears that the marsupials evolved before the placental mammals. They first became important about 100 million years ago and had spread to all the continents by 70 million years ago, which implies that all the world's landmasses were interconnected at some point during this time. Within a few million years of the later date, placental mammals evolved and swept across Europe, Asia and North America, replacing the marsupials in those areas. In South America, however, the marsupials continued to occupy their ecological niche until about 40–30 million years ago, when they were finally replaced by the placentals. The inference seems clear: namely, South America must have been isolated from the northern continents some time after 70 million years ago but reconnected after 40 million years ago. Meanwhile, Australia must have been separated at about the same time as South America but never reconnected, since, when Cook landed in 1770, the mammal population consisted entirely of the marsupial variety.

In the past this evidence has been interpreted as showing that huge areas of land once connected the continents but sank beneath the waves during some cataclysmic upheaval of the Earth. There is indeed a very long history to legends of the 'Atlantis' variety, but today continental drift seems a far simpler way of explaining the observations.

2.2.5 Palaeomagnetism

The piece of evidence that was not available to Wegener but that has been most instrumental in persuading scientific opinion to Wegener's views is that provided by the study of ancient rock magnetism.

The reader will probably be familiar with the idea that the Earth acts rather like a gigantic bar magnet with a dipole field (Fig. 2.4). The north magnetic pole does not correspond exactly with the north rotational pole, being centred in northern Canada at the moment. At the present day the magnetic field can be mapped using a compass, but it has proved possible to determine the magnetic field of the past from evidence contained within rocks of suitable composition. Some lavas (e.g. basalt, Appendix A.2) contain a high proportion of iron-rich minerals. Basalt usually erupts from a volcano in liquid form and solidifies as it cools. Between about 600 and 500 °C the geomagnetic field tends to induce magnetism within the iron-rich minerals in line with the prevailing magnetic field. As further cooling takes place, this induced magnetism becomes 'frozen'; it will persist indefinitely unless the rock is reheated back to 600 °C. Careful laboratory analysis of a sample of the rock after cooling allows determination of two important properties:

(a) The alignment of the geomagnetic field prevailing at the time of cooling, relative to the Earth's present magnetic pole.
(b) The latitude of the sample, relative to the magnetic pole prevailing at the time of cooling.

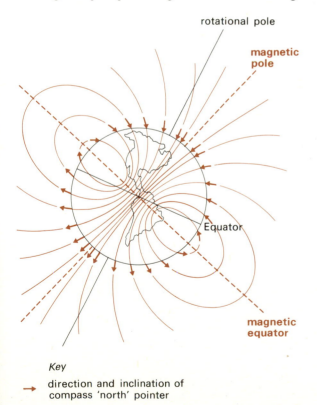

Key

→   direction and inclination of compass 'north' pointer

Figure 2.4 The Earth's magnetic field, showing the changing inclination of a compass needle with latitude.

This can be determined by the inclination of the magnetic field to the horizontal. As Figure 2.4 indicates, the geomagnetic field lies parallel to the Earth's surface at the magnetic equator and enters the surface at an increasingly high angle as the magnetic pole is approached. A freely suspended magnet (or an induced magnetism in a rock sample) will therefore be horizontal at the magnetic equator and vertical at the magnetic pole, and at positions between the relationship is determined by

$$\tan I = 2 \tan L$$

where I is the angle of inclination from the horizontal and L is the angle of magnetic latitude.

In the last few years, this analysis has been carried out for lavas, and in some cases sedimentary sandstones, of widely varying age. If all the palaeomagnetic data for, let us say, Africa are put together, it is possible to plot the results on a map. In Figure 2.5a the coloured line superimposed over Africa shows the *apparent* position of the south magnetic pole relative to the continent over time. This line is called a **polar-wandering curve**, and it may be interpreted in one of two ways:

(a) as showing that the magnetic pole has changed position on the Earth's surface; or
(b) as showing that the continent has moved relative to the magnetic pole.

As far as the first is concerned, it is already known that the magnetic pole changes its position on the Earth's surface; the fact is recorded on the margin of any Ordnance Survey map, where true and magnetic north are indicated. But that variation is short term and small scale and, over any length of time, it seems that the average magnetic-pole position is the same as the rotational pole position, which is not very surprising, since it is the rotation of the Earth that produces the magnetic field in the first place. There is no other evidence to suggest that the magnetic pole may have wandered across the rotational Equator as is indicated in Figure 2.5a. Closer examination of that diagram reveals a further problem. The coloured line drawn through the South Atlantic is the polar-wandering curve for South America. It joins the polar-wandering curve for Africa as the modern magnetic south pole is approached, but northwards (and backwards in time) the two curves diverge. If this is to be interpreted purely in terms of magnetic field movement, clearly there must have been two magnetic south poles – a most unlikely situation. A far easier explanation is that the two continents have wandered across the Earth's surface. About 400 million years ago, both lay on the other side of the South Pole. Between 400 and 300 million years ago Africa lay over the pole, and since that time both have been moving to the north. The diverging curves can be explained by assuming that the two continents were once joined together but have split apart. Indeed, it is possible to take the two halves of Figure 2.5a and overlay the two polar-wandering curves (taking due care over the map projection problem). As Figure 2.5b shows, if this is done, the continents fit together in precisely the same manner as suggested by shape and structural considerations.

This forms a most convincing piece of evidence for continental drift, and the fact that the same exercise can be carried out for the North Atlantic rather tends to clinch the matter. Certainly, the overwhelming bulk of scientific opinion now accepts that the continents have not remained fixed in their present positions throughout geological time.

2.3 The evolution of the present pattern of continents

By using all the evidence of the preceding section, especially the palaeomagnetic data, in combination with the fairly accurate dating of rocks using radio-isotope methods (Appendix B), it is possible to give some idea of how the present continental pattern has evolved.

The Earth itself is considered to be approximately 4550 million years old. This figure is derived largely from the age of Moon rocks and meteorites, since the oldest rocks visible at the surface of the Earth are closer to 3800 million years. This latter figure suggests the age of the continents themselves, since the oldest rocks are found in the most stable areas of the continental interiors – the so-called 'shield' areas of Canada, Brazil, the Baltic, Siberia, the Sahara, southern Africa, India and Australia. These core elements, often referred to as **cratons**, were probably once part of one proto-continent, which repeatedly split up and regrouped, forming new shape combinations. Any one continent would be shaped partly by the line of the split itself (Ch. 4)

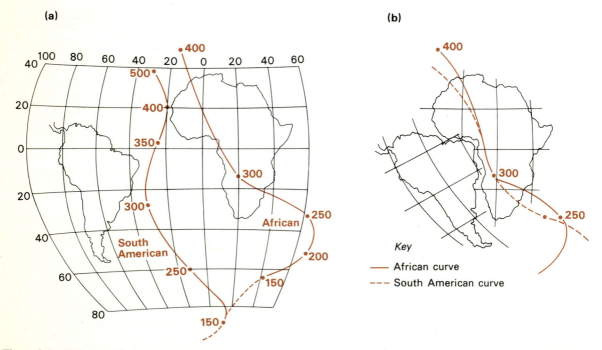

Figure 2.5 Polar-wandering curves. (a) The apparent position of the south magnetic pole relative to Africa and South America for various dates in the past (figures in million years ago). (b) The superimposition of the two polar-wandering curves produces a continental fit identical to that produced by shape alone (from Tarling & Tarling 1972). Note that in (b) the shape of South America and its polar-wandering curve have been redrawn using the same projection as in (a) but centring the continent over the projection meridian.

and partly by the sedimentary plains or mountains continuously being added to the continental margins (Chs 5 and 6). The last occasion on which all the continents seem to have been grouped together was about 200 million years ago (Fig. 2.6a). The 'super-continent' so formed has been called Pangaea and was, as we have already seen, the setting for tropical swamps along the line of the Equator, which eventually provided the coalfields for much of the industrial world. At the same time the southern hemisphere was experiencing a glacial epoch, with ice sheets covering a large area of present-day Antarctica as well as the adjacent landmasses. Over time Pangaea split up again. At first the line of splitting lay east–west, to give two subunits: Laurasia in the north, comprising North America,

Greenland, Europe and most of Asia; and Gondwana in the south, made up of South America, Africa, India, Australia and Antarctica. At first the two were joined by a land bridge connecting North Africa with Laurasia, but this link became more tenuous as the earlier Tethys Sea opened along the line of the Mediterranean and the Caribbean was created by the rifting of the Americas. At first Laurasia drifted northwards (Fig. 2.6b), and the British Isles moved from tropical swamp to desert climate. Increasingly, the two continents subdivided. In the south, India seems to have moved quite rapidly to the north, Africa and South America moved more slowly, while Antarctica and Australia remained stationary for some time. Laurasia was also splitting apart as the Atlantic

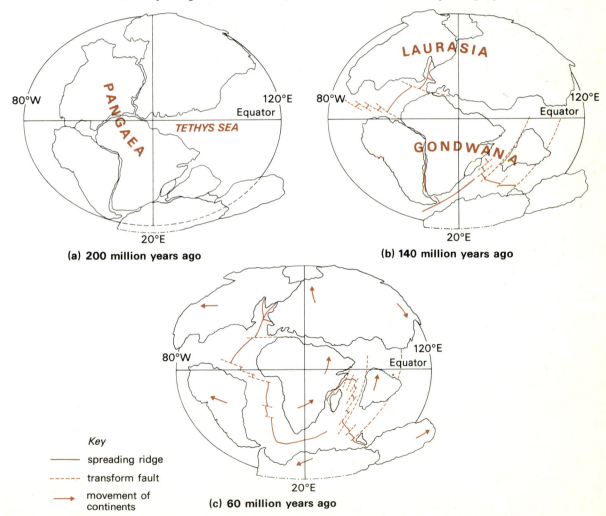

(a) 200 million years ago

(b) 140 million years ago

Key
——— spreading ridge
------- transform fault
——→ movement of continents

(c) 60 million years ago

Figure 2.6 A reconstruction of the position of the continents at three times during the last 200 million years (from Dietz & Holden 1970).

came into being, with rifts separating either side of Greenland.

As late as 70 million years ago, the gaps between the continents were probably still quite narrow, and a certain amount of movement could still take place among land animals. By 50 million years (Fig. 2.6c), however, the continents were moving to their present positions and regrouping into new patterns. Having drifted mainly westwards, the two parts of the Americas also rotated until they touched at Central America. The continued northward movement of Africa and India closed most of what remained of the Tethys Sea, while the continued opening of the Atlantic eventually separated Europe from North America.

2.4 Continental drift and climatic change

A subject that at first may seem unrelated to the present discussion but is worth introducing at this point is the whole question of climatic change. In one sense, the preceding section should make it clear why the climate of any one area of the world may change. If a continent is drifting across the surface of the Earth, over time it will pass through a number of climatic zones. Thus, 300 million years ago the British Isles were experiencing equatorial conditions, 250 million years ago the climate of a hot desert, and so on.

A more complex subject is changes in the global climate. A considerable amount of work has been carried out in the last few years on the causes of climatic change on this scale. Quite apart from man's unintentional interventions, short-term variations may be due to changes in solar radiation, but longer-term variations (e.g. those responsible for the waxing and waning of the ice sheets over the last 2 million years) are probably attributable to regular alterations in the orbit and axial tilt of the Earth, none of which has anything to do with continental drift. All the changes that occurred during the last 2 million years have, however, taken place within what is generally called an **ice age**. This is a long period of low average temperature, which may include numerous **glacial** episodes, characterised by severe cold and ice sheet growth, interrupted by **interglacials**, during which world temperatures may be considerably above those of the present day. An ice age is in operation at the moment and has been for the last 2 or 3 million years. We are living during an interglacial, which was preceded by perhaps as many as fourteen separate glacials since the start of the current ice age. Ice ages in general seem to be

fairly rare occurrences in the history of the Earth. Before the present one there were at least two others, at approximately 300 and 600 million years ago. This is far too long an interval to be caused by variations in solar radiation, orbit or axial tilt; what is required is a much longer-term change. One possible change is, of course, continental drift.

Any basic climatology course teaches that landmasses warm up and cool down more quickly than the sea. The sea is said to have a moderating effect upon temperatures, whereas 'continental' areas have extreme temperature ranges. If, during one period of Earth history, the polar regions are occupied by oceans, ice will be more limited in extent, since continuous water-mixing will prevent surface temperatures' dropping too low. A concentration of landmasses at the poles, however, should have the opposite effect, encouraging the growth of ice and the initiation of an ice age. During the Carboniferous period (300 million years ago) there was a concentration of landmasses at the South Pole and a corresponding ice age. Since that time, as Figure 2.6 indicates, although Antarctica has remained more or less stationary, there has been a general movement away from the poles until the last few million years, when North America and Eurasia have surrounded the North Pole. Although the North Pole itself is not a land area, the Arctic Ocean is virtually landlocked, and water circulation is therefore limited.

None of the foregoing discussion offers 'proof' of the role of continental drift in the initiation of ice ages, and other mechanisms may well be at work. Nonetheless, continental drift and the land–sea pattern that it produces may at least be a contributory factor in the alteration of global climate.

2.5 Conclusion

In this chapter we have shown that there is now an overwhelming body of evidence in favour of continental drift, and indeed it is possible to write a history of continental drift or even discuss some possible consequences of drift. What we have, of course, studiously avoided up to this point is any discussion of the *cause* of continental drift. No matter how good the evidence for something like continental drift, scientific opinion is almost bound to remain somewhat sceptical if no reasonable mechanism is advanced to explain how the whole scheme works. It is precisely to this problem that much attention has been given in recent years, and it is therefore necessary for us now to examine what is known about the interior workings of the Earth.

Chapter 3

The Internal Structure and Mechanisms of the Earth

3.1 The bulk composition of the Earth

Over the years many theories have been advanced to explain how the Earth, and the other planets in the solar system, originated. Many writers have imagined that the planets formed as the result of some catastrophic event, such as a collision between stars or the close approach of another star, causing material to be drawn from the Sun by gravitational pull. However, the fact that all the planets rotate around the Sun within the same plane of orbit, and in the same direction as the spin of the Sun itself, strongly suggests that both Sun and planets had their origin in a rotating disc of material. Originally, it was thought that the rotating disc consisted of hot gas, but opinion now generally favours a disc of cold dust and gas. The material of that disc was dominated by the lighter elements, hydrogen and helium. Gravitational forces gradually concentrated much of the material into the centre, where increasing pressure raised the temperature to the point where nuclear fusion processes could start, and the Sun was born. The rest of the material in the disc, although still held to the centre by gravity, was also subject to the outward (centrifugal) push caused by rotation. The lighter elements not absorbed by the Sun seem to have concentrated towards the outer edge of the disc, where they coalesced to form the 'gas giants': Jupiter, Saturn, Uranus and Neptune. The Sun, with a density of 1.4 g/cm³, lies within the density range of the gas giants (0.7–1.7 g/cm³). The small amount of heavier elements in the solar system seems to have become concentrated in the area between the Sun and Jupiter, where it subsequently formed the inner planets: Mercury, Venus, Earth and Mars. All these planets are much smaller than the gas giants but have much higher densities, the range being from 4.0 g/cm³ for Mars to 5.5 g/cm³ for Earth.

The total density of a planet (e.g. Earth) is calculated using astronomical data of size and distance within the laws of gravity in order to determine mass. The figure is therefore independent of any surface observation but clearly leads us to expect a dominance of heavy elements in the Earth. In fact, the average composition of the surface of the continents includes 47 per cent by weight of oxygen and 28 per cent by weight of silicon. Although these two elements are clearly much heavier than hydrogen and helium, they are not heavy enough to produce the observed average density of 5.5 g/cm³ if present throughout the Earth in this quantity. One is led to suspect, therefore, that within the Earth the balance of elements is not the same as at the surface. This view is reinforced by the analysis of meteorites, which are thought to be fragmentary planetary material. A small proportion of all meteorites is composed entirely of an iron–nickel mixture, and this has led observers to suggest that part of the Earth's interior may have a similar composition.

There is therefore some evidence for chemical variations beneath the Earth's surface. There is also, of course, some very compelling evidence for the variations in physical conditions that exist, since the eruption of a volcano implies that rock temperatures below ground are considerably higher than those at the surface.

The Earth has an average radius of 6371 km. The word 'average' has to be used, incidentally, since the Earth is not a true sphere but is somewhat flattened at the poles (an oblate spheroid!). The deepest mines and boreholes cannot possibly extend for more than a tiny fraction of that distance, and we cannot therefore directly examine the interior of the Earth to determine how chemical and physical properties vary. Instead, we have to use indirect methods of determination, and of these by far the most useful is the analysis of earthquake waves (**seismology**).

3.2   Seismic investigations of the Earth's interior

The cause, location and effects of earthquakes will be discussed at several other points in this book, so that for the moment it is sufficient to know that an earthquake is the result of earth movements. If stress is placed upon a section of rock by forces operating in opposite directions (Fig. 3.1a), the rock initially deforms, because it is to some extent elastic (Fig. 3.1b). If stress is maintained for long enough, the strain may exceed the elastic strength of the rock, and the rock will fracture along the line of a **fault**. At the moment of fracture the rock regains its original shape but in a new position (Fig. 3.1c). It is during the sudden movement of the rock back to its original shape, after the stress is released, that the ground 'shakes'. The 'shaking' results in the propagation of waves, which move through rock somewhat as waves spread out on a pond from a stone dropped into the water. The waves emanate from a point, the **focus**, which may be some depth beneath the ground. As far as observers at the surface are concerned, waves appear to emanate from a point on the ground surface that is vertically above the focus; this is called the **epicentre** (Fig. 3.3).

The damage produced by earthquakes is due to the three-dimensional movement of the ground as surface waves travel out from the epicentre (see Ch. 7). Seismic analysis, however, employs waves that travel beneath the surface, and these are of two types:

(a) a compressional wave; and
(b) a distortional wave.

Figure 3.2 is an attempt to provide a visual impression of these waves in terms of their passage through a piece of rock.

As with waves of other types (e.g. sound waves), earthquake waves have a finite velocity, but compressional waves will always travel faster than distortional waves, and so they have come to be renamed **primary** and **secondary waves** respectively (P-waves and S-waves).

At many points on the globe are observation stations equipped with **seismographs**, which record the passage of seismic waves. Careful analysis of a seismograph record allows different types of waves to be identified. For any particular seismic event it may be possible to pick up P-waves and S-waves at varying distances from the epicentre. In Figure 3.3a, for example, an event with its epicentre at X is identified at stations A, B and C. When plotted on a graph (Fig. 3.3b) a simple relationship exists between the distance from the epicentre and the time taken for seismic waves to arrive. Notice that the distance scale of this graph is plotted in degrees of arc rather than surface measurement. This is done purely for convenience, since it is useful to think in

(a) **(b)** **(c)**

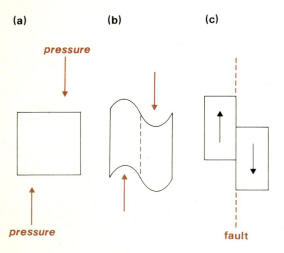

Figure 3.1   Diagram illustrating the immediate cause of an earthquake. (a) A section of rock is placed under stress acting from opposite directions. (b) At first the rock deforms. (c) Beyond a certain point the rock fractures, and the rock splits in two along a fault, each part rebounding back to its original shape but at a new position. It is the rebounding that causes an earthquake.

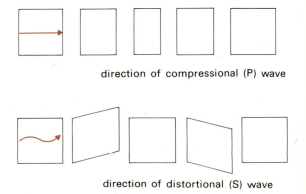

direction of compressional (P) wave

direction of distortional (S) wave

Figure 3.2   Graphical illustration of the differing effects of compressional and distortional waves on rock material.

terms of parts of a sphere, but it amounts to exactly the same thing in the end.

In Figure 3.3b there is a linear relationship

(a)

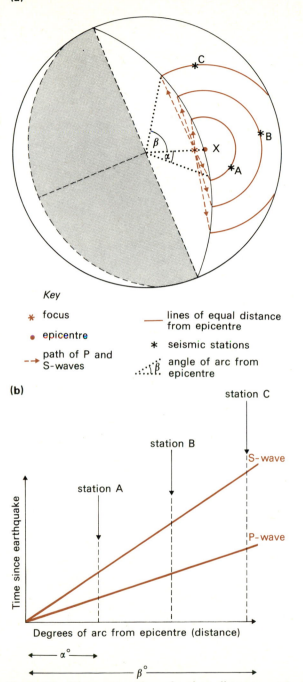

Key

* focus

• epicentre

→ path of P and S-waves

— lines of equal distance from epicentre

* seismic stations

∴ β̣ angle of arc from epicentre

(b)

Figure 3.3 The construction of a time–distance graph for an earthquake event (b) from data collected at three seismograph stations (a).

between time and distance, which implies that the velocity of the waves is constant between epicentre and recording station. In practice, the velocity of an earthquake wave may vary, depending upon a number of factors. In particular, waves tend to travel faster through denser rock, but they are slowed down if there is a change from solid to liquid rock conditions (in fact, S-waves will not penetrate liquid materials at all). Figure 3.4 shows a slightly more realistic time–distance graph for the seismic event shown in Figure 3.3a. In this case the P- and S-waves reach the more distant seismic stations more quickly than might be expected; in other words, they have a higher average velocity than waves reaching the closer stations. Examination of Figure 3.3a shows that waves travelling to the farther stations pass through rocks at greater depth than those travelling to the nearby stations. One explanation of Figure 3.4 therefore is that rocks at depth have a greater density, causing earthquake wave velocity to increase with depth. Broadly speaking, this conclusion holds true for much of the Earth and demonstrates that rock density increases beneath the surface, although there are some very important exceptions to this principle.

So far we have only considered waves that travel directly from focus to recording station. Examination of time–distance graphs for somewhat greater distances shows a rather curious pattern. Figure 3.5a is a much simplified time–distance

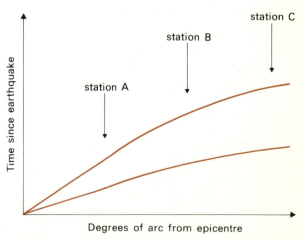

Figure 3.4 A time–distance graph for a single earthquake event, showing the convex line produced when waves travelling through deeper layers of the Earth attain a higher average velocity due to increasing rock density with depth.

graph covering 180° of arc (i.e. an entire hemisphere). The first part of the graph looks like the example that we have already examined, except that there appear to be two P-waves and two S-waves recorded at each station. Further round, the situation is even more strange, for no P-waves at all are recorded between about 105° and 143°, but they reappear after that point. The S-waves apparently stop entirely at 105°.

This complex pattern is interpreted as showing that deep within the Earth there is a break, or **discontinuity**, between two distinct layers. The imagined situation is shown in Figure 3.5b. The lines on graph and section marked DP show the passage of **direct** primary waves, which have already been discussed. Beyond 105° direct primary waves cannot reach the ground surface without crossing the discontinuity. When a wave meets a discon-

tinuity in the Earth, it will act somewhat like light waves on striking a water surface. On the one hand, the wave is **reflected** back off the discontinuity, the angle of reflection being (as in optics) equal to the angle of direct wave incidence (Fig. 3.5c). Lines marked RcP in Figure 3.5 show the behaviour and recording of primary waves reflected from the discontinuity and explain the apparent 'double' wave recorded at stations up to 105°. At the same time the wave crosses the discontinuity into the lower layer. If the velocity of wave travel is different in the two layers, the path of the wave is bent, or **refracted**, as it passes across the discontinuity (Fig. 3.5c). The angle of refraction will depend upon the relative velocity of wave travel in the two layers; but from our point of view the only thing that matters is that, if the wave velocity of the lower layer is higher, the wave will be bent away from the centre of the

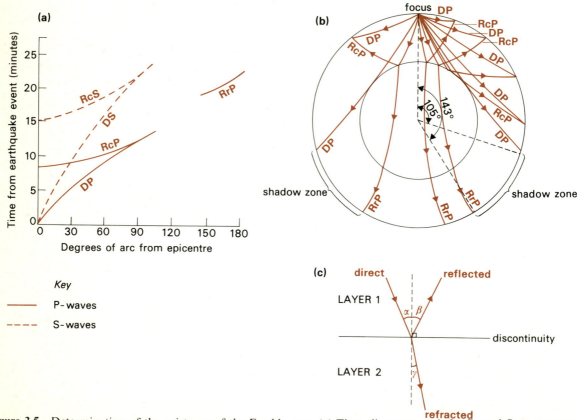

Figure 3.5 Determination of the existence of the Earth's core. (a) Time–distance graph for P- and S-waves across a hemisphere, showing direct, reflected and refracted wave patterns as well as a shadow zone. (b) Diagrammatic section through the Earth, showing how the presence of a core with lower seismic wave velocity would produce the wave pattern shown in (a). (c) Definition of direct, reflected and refracted seismic waves at a discontinuity. The angle of incidence (α) is the same as the angle of reflection (β), but the angle of refraction (γ) depends upon the relative velocity of seismic waves in layers 1 and 2. As illustrated, layer 2 has a lower velocity.

Earth, whereas it will be bent towards the centre if the lower layer has the lower velocity. As Figure 3.5 suggests, the only reasonable interpretation of the absence of P-waves between 105° and 143°, the **shadow zone**, is that P-waves are refracted towards the centre of the Earth (lines marked RrP). The conclusion that this inner part of the Earth has a lower seismic wave velocity is a little curious, since one might expect rock density to increase with depth. The fact that S-waves do not seem to cross the discontinuity at all, however, suggests that this lower region may be liquid, and that would also be sufficient reason to account for the slowing down of P-waves.

One final point on Figure 3.5: the paths of earthquake waves on the section (Fig. 3.5b) are shown as curves rather than straight lines. This is because continuous changes in wave velocity with increasing rock density have the effect of causing continuous refraction or bending of the paths. In all cases within each layer, the path is curved away from the centre of the Earth as velocity increases downwards.

Up to this point we have been able to show that the analysis of seismic waves reveals a good deal about continuous changes in rock density as well as sudden changes, or discontinuities, within the Earth. Already it has become apparent that the Earth consists of at least two layers, with the inner layer being at least partly liquid and both layers showing increasing density with depth. Further analysis of this sort is well beyond the scope of this book, although the principles remain the same, and that should be borne in mind as we examine the results of seismic investigation in the next section.

3.3   The internal structure of the Earth

The simplest way of representing a good deal of the seismic evidence available is to draw velocity profiles for the Earth showing seismic wave velocity at varying depths. Figure 3.6 is an attempt to show the velocity profile of the Earth to its centre, and Figure 3.7 is a more detailed profile of the top 50 km.

The most marked feature in Figure 3.6 is the major discontinuity at 2900 km below the surface. This discontinuity effectively divides the Earth into the two layers identified in Figure 3.5b and represents a sudden change from solid to liquid conditions, shown in the dramatic slowing of P-waves and the halting of S-waves. Below this discontinuity is the **core** of the Earth, which has an approximate radius of 3500 km. The core has a very high density,

which tends to suggest that it is made largely of iron (probably 91 per cent by weight) plus nickel. The pattern of P-wave velocity within the core (Fig. 3.6) indicates that it may be divided into an **inner core** (density 14–16 g/cm³, temperature around 4000°C, which acts as a solid, and a surrounding **outer core** (density 9.7–11.8 g/cm³, temperature around 3500°C), which acts as a liquid.

The outer layer of the Earth, lying between 10–60 and 2900 km below the surface, is called the **mantle**. In contrast to the core, the mantle is at a lower pressure, temperature (1000–3500°C) and density (3.3–5.7 g/cm³). It too may be subdivided into **lower** and **upper mantle** (Fig. 3.6) on the basis of the velocity profile. The lower mantle is thought to be made largely of oxides of magnesium, iron and silicon, which, being under high temperature and pressure, have no direct counterparts in surface rocks. The upper mantle, however, consists of minerals that can also be found at the surface. The commonest are probably the silicate minerals olivine and pyroxene (Appendix A.1), which together form the rock called **peridotite** (Appendix A.2). Additional minerals like the silicate **garnet** and the iron oxide **spinel** are probably present.

In the early days of seismic wave analysis a clear discontinuity was identified quite close to the surface of the Earth. Named after its discoverer, the

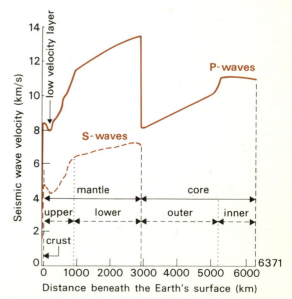

Figure 3.6   The velocity profile of the Earth, based on the averaged travel velocity of P- and S-waves with increasing depth. Marked breaks in the profile define the main layers within the Earth. The low velocity layer at 100–250 km stands out clearly (after Clark 1971).

Mohorovičić discontinuity represents the junction between the hot and solid, but slowly deforming, material of the mantle and the cold and rigid material of the Earth's **crust**. The most interesting thing about the crust is that it shows some marked variations. Over the oceanic areas of the world (Fig. 1.1) the crust is thin, being less than 10 km thick (Fig. 3.7) and remarkably uniform in its composition. The rocks of the oceanic crust are rather richer in aluminium and calcium than the mantle but also include iron, magnesium and silicon to form the minerals olivine, pyroxene and anorthite feldspar (Appendix A.1). Together, these produce the rock **gabbro**, or its fine-grained equivalent, **basalt** (Appendix A.2).

Beneath the continental areas, including the continental shelves, however, the crust is substantially thicker, being about 33 km thick on average but increasing to over 65 km thick beneath mountain ranges. The density of continental crust is lower than its oceanic equivalent (2.7 g/cm³ as against 2.9 g/cm³), and this is a reflection of chemical differences. Continental crust is richer in silicon,

sodium and potassium and poorer in iron, magnesium and calcium. This means that minerals such as quartz, mica and alkali feldspar (Appendix A.1) are more common in continental crust. The other important difference is that continental crust shows an enormous diversity in rock type and structure and cannot be said to consist simply of one rock type. Nonetheless, it would not be entirely unjustified to indicate that, beneath the surface veneer of sedimentary rocks that covers large areas of the continental surfaces, continental crust is dominated by igneous rocks such as **granite** and **granodiorite** (Appendix A.2) and metamorphic rocks rich in quartz and feldspar (Appendix A.5). Finally, it should be added that there is something of an age difference between oceanic and continental crust. Although continental areas include rocks from virtually every geological age (Appendix B), substantial areas are very old, dating well back into the Precambrian (more than 600 million years old). No part of the oceanic crust, however, is more than 200

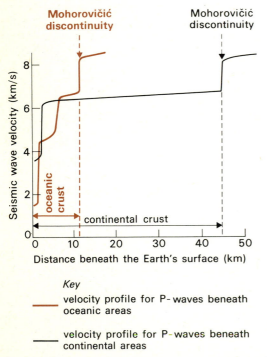

Figure 3.7 The velocity profile for the top 50 km of the Earth, using P-wave data only. Separate profiles are shown for depths beneath oceanic and continental areas to illustrate the different depths at which the Mohorovičić discontinuity occurs (after Clark 1971).

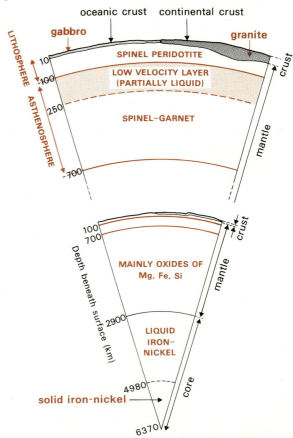

Figure 3.8 Generalised segment of the Earth at two scales to show the main layers beneath the surface.

million years old, and a good deal is much younger than that.

In recent years considerable attention has been given to the crust and upper mantle, and some doubt has been raised about the fundamental importance of the Mohorovičić discontinuity. Further examination of Figure 3.6 will show that seismic wave velocity falls with depth over part of the upper mantle. This **low velocity layer**, lying between 100 and 250 km below the surface, appears to represent a layer within the upper mantle where the rocks are at least partly liquid. Generally, the mantle rocks are solid, although, as we have already indicated, they will deform slowly. A partially liquid layer, however, would allow much greater movement of material, and it begins to look as though the rocks within that layer may be capable of shearing, thus allowing the upper mantle and crust above to move over the remaining mantle below. In recognition of this possibility, the uppermost mantle and crust are now referred to as the **asthenosphere** and **lithosphere**. The lithosphere is the top 100 km or so of the Earth's interior, consisting of crust plus rigid mantle rocks beneath. The asthenosphere is that part of the upper mantle comprising the low velocity layer and more viscous mantle rocks beneath, down to about 700 km. Figure 3.8 is an attempt to show these layers visually.

3.4   Heat flow and convection within the mantle

Traditionally, it has been assumed that the Earth is hot inside, because the whole planet was once molten and has subsequently cooled down, at least on the outside, where the 'crust' formed. As indicated earlier, however, opinion now favours a cold start to the solar system. This presents no difficulty as far as the Sun is concerned, since gravitational forces acting within that enormous mass can produce the temperatures required to initiate nuclear fusion, during which hydrogen is converted to helium and energy is given off. The centre of the Earth at 4000 °C is much too cold for fusion to take place. On the other hand, nuclear fission processes, involving the breakdown of heavy isotopes, do not require high temperatures. Without doubt the breakdown of radioactive uranium and thorium does occur within the Earth, and this does produce heating, but both these isotopes are rather rare. The radio-isotope of potassium (^{40}K), however, is fairly abundant and also breaks down, giving off energy. It seems likely that a combination of fission processes is responsible for maintaining the high temperatures within the Earth. Do not imagine, though, that the Earth is gradually heating up over time! During the course of millions of years the heat generated internally is balanced by an

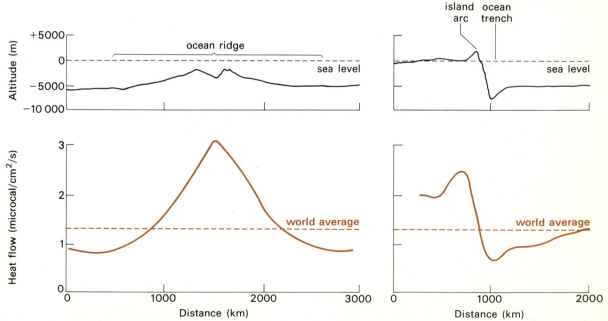

Figure 3.9   Heat flow variations. Topographic sections across a typical ocean ridge (a) and island arc–ocean trench (b) are shown above corresponding graphs of averaged heat flow (partly from Tarling & Tarling 1972).

equivalent loss of heat through the surface of the Earth and out into space. On the other hand, the Earth probably had radio-isotopes in greater abundance in the past, and so in that sense the planet is cooling down.

The loss of internally generated heat through the surface of the Earth is called **heat flow**. Heat flow may be measured at the surface, but the quantities concerned, although enormous on a world scale, are scarcely detectable. On average the amount of energy coming up through the ground is only 1/10 000th of the energy being received at the same spot by solar radiation (and so houses in this country still need central heating!). Much of the energy loss measured as heat flow is being transferred from core and mantle and out through the crust by direct **conduction** (the same process that causes the handle of a saucepan to become hot, even though it is not in direct contact with the source of heat). There must be some other heat transfer processes also at work, however, because considerable variations in the value of surface heat flow exist. As Figure 3.9 indicates, higher than average heat flow is recorded across ocean ridges and island arcs, whereas deep ocean trenches have decidedly low heat-flow values. This suggests that there are areas of higher and lower average temperature

within the mantle at these locations. In most cases the high heat-flow values coincide with the occurrence of volcanoes, which are, after all, simply the surface expression of high subsurface temperatures. A map of volcano location (Fig. 3.10) in fact makes a good starting point for looking at the heat flow pattern. Although some volcanoes (e.g. those on Hawaii) occur in isolated groups, the majority of the world's volcanoes lie along linear belts, which correspond closely to the linear features noted in Figure 1.1. It is possible to identify at least four linear groupings:

(a) *ocean ridges* (e.g. the Atlantic ridge), along which occur volcanic islands (e.g. Iceland, the Azores);
(b) *continental rift valleys* (e.g. that in East Africa);
(c) *volcanic island arcs* (e.g. the Philippines, the Marianas), which usually lie parallel to deep ocean trenches; and
(d) *major mountain chains*, some of which (e.g. the Andes) are coastal and lie parallel to ocean trenches, whereas others (e.g. the Alpine–Himalayan chain) are within continental masses. Both types of mountain chain support volcanoes.

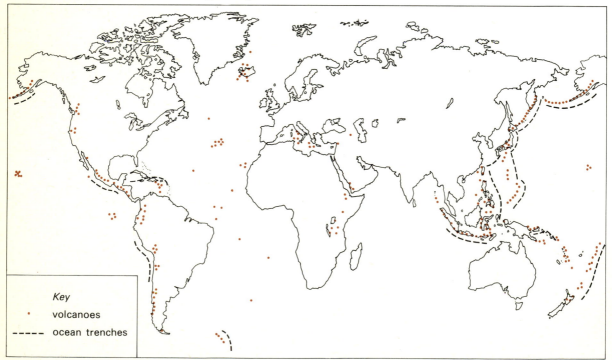

Key
· volcanoes
----- ocean trenches

Figure 3.10   The world distribution of active volcanoes, shown together with the major deep ocean trenches.

Ocean ridges and rift valleys are areas only of high heat flow, and it therefore seems likely that they lie above zones in the mantle where temperatures are higher than average. Many observers consider that this situation is created by rising plumes of hot mantle rock. This may be a transfer of heat by some process similar to **convection**, which normally only occurs in gas or liquid. If a 'parcel' of gas or liquid is heated, it will expand, with the result that its density decreases. Being less dense than its surroundings, it will rise relative to the rest of the gas or liquid body (this process is clearly seen in the rising of hot air above a heater in a room). So far we have suggested that most of the Earth's mantle, although fairly hot, is still solid. It is possible, however, for some solids to deform slowly when under stress, a simple example being shown by the glass in old window panes, which is invariably thicker at the bottom of the pane as a result of flow. The mantle rocks may behave in a similar fashion and rise towards the surface under ocean ridges and rift valleys, transferring heat in the process.

If mantle rocks are rising towards the surface of the Earth in some places, there must be a counter-flow of rock back into the mantle somewhere else to maintain continuity of matter. Since that counter-flow must presumably involve cooler rock from near the surface, the site of a descending flow should be marked by low heat-flow values at the surface. The deep ocean trenches presumably mark the descending flow, although this does not immediately explain why island arcs are usually found next to trenches.

Put together, then, we have a simple model of convection currents within the mantle, rising along ocean ridges and descending at ocean trenches.

3.5 Plate tectonics

In Chapter 2 we saw some fairly convincing evidence in favour of continental drift. In this chapter we have seen that something like a convection system exists within the mantle as a result of heat flow. The question obviously now arises: does heat flow provide the mechanism that *causes* continental drift?

When this possibility was first considered, it was envisaged that the continents would move around

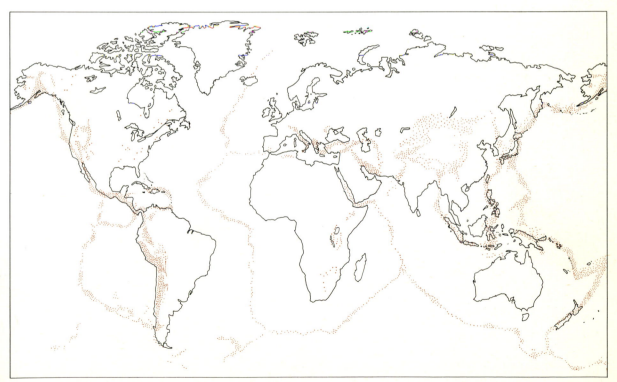

Figure 3.11 The world distribution of earthquake activity (from US Coast and Geodetic Survey; after Clark 1971).

the surface of the Earth 'floating' on mantle rocks almost like rafts on a lake. Since that time it has become clear that there are many objections to this idea. For the moment, think about what must happen at the boundaries of the continent. First, as far as thickness is concerned, if the continent were moving as an isolated unit, movement would necessarily take place at the crust/mantle boundary (i.e. the Mohorovičić discontinuity). Although there is evidence of a clear physical change at this boundary, there is nothing to suggest that shearing or flow occurs here. Second, as far as surface area is concerned, if the continent were moving across an underlying layer, we would expect continuous faulting and earth movement to occur along the margins of the continent. In fact, if we use a map of the world distribution of earthquakes (Fig. 3.11) as a guide to active faulting, it immediately becomes clear that continental margins do not correspond very well with active fault zones.

Both these difficulties can be resolved, to a large extent, if we assume that the continents do not move in isolation but as parts of larger moving units. For example, although there is no evidence of

movement at the Mohorovičić discontinuity, the asthenosphere at greater depth is thought to be semi-liquid, which would provide a layer within which shearing or flow may occur. In other words, it is quite possible that the whole lithosphere (i.e. the crust and top part of the mantle) moves as a single layer over the asthenosphere.

In the same way, consider the likely surface area of movement. To reverse an earlier argument, does the map of world earthquake distribution (Fig. 3.11) define the areas that are moving? From this map alone we can imagine the world as consisting of a number of irregularly shaped pieces (Fig. 3.12), like the bits of a jigsaw puzzle. Comparison of the earthquake map with the volcano map (Fig. 3.10) shows that earthquakes are generally related to areas of high or low heat flow, and it is very tempting to interpret this relationship as showing that the convection currents of mantle heat flow are causing the surface movements that generate earthquakes.

Does it follow that such a straightforward relationship could account for movement of the lithosphere across the Earth's surface, which would

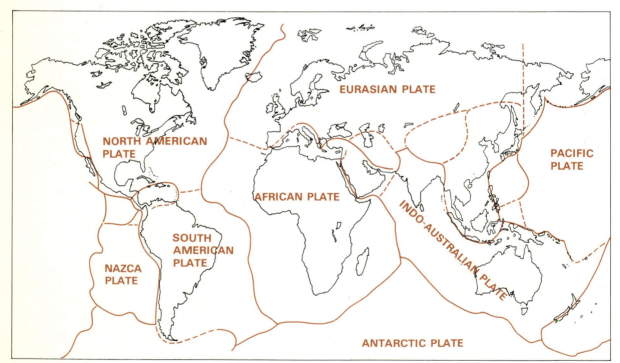

Figure 3.12 A possible division of the Earth's surface into rigid plates, based largely upon seismic data. Solid lines indicate boundaries where other evidence supports the idea of a plate boundary (see Chs. 4 and 6). Dotted lines indicate uncertain boundaries.

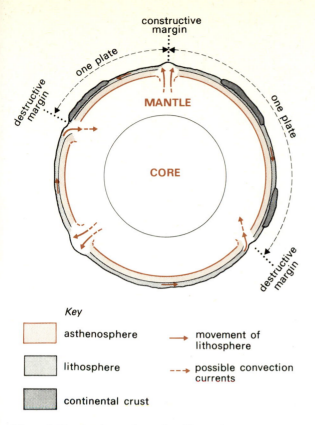

constructive
margin

one plate

destructive
margin

MANTLE

one plate

CORE

destructive
margin

Key

asthenosphere

lithosphere

continental crust

→ movement of
lithosphere

--→ possible convection
currents

Figure 3.13 A schematic section illustrating the mechanism envisaged by plate tectonics. Rising currents within the mantle create lithospheric material at a constructive margin. Lithospheric plates move away from constructive margins, sliding over the underlying asthenosphere. Lithospheric material is destroyed at destructive margins, which lie above descending mantle currents.

produce continental drift? One immediate problem arises. If the entire surface of the Earth is covered by pieces of lithosphere, how is there room for movement to take place? One possibility is that the pieces of lithosphere on the Earth's surface act something like the moving top of a conveyor belt. The active margins of each moving unit (i.e. the earthquake belts) represent the ends of the conveyor belt, where lithosphere is either moving up to the surface or moving down into the interior of the Earth. As Figure 3.13 suggests, the upwardmoving edges would be powered by rising convection currents, which would have the effect of creating new lithosphere at the surface, whereas the downward movement would correspond with sinking convection currents and would represent the remelting of lithosphere. All the evidence that we have so far accumulated on heat flow and surface topography tends to support such a model.

The model just outlined represents the bare bones of the elaborate, although as yet incomplete, set of ideas normally referred to as **plate tectonics**, the word 'plate' being used to describe a single moving unit of lithosphere. In the next three chapters, we are going to examine plate tectonics in more detail, by looking at the active margins and surfaces of the plates defined in this section. We shall start by looking at the evidence for the creation of lithosphere in some places (Ch. 4, 'Constructive plate margins'), and we shall end with the evidence for the destruction of lithosphere in other places (Ch. 6, 'Destructive plate margins').

Chapter 4

Constructive Plate Margins

From the evidence examined so far, it seems that some of the tectonically active belts of the world correspond to areas of high heat flow. These belts may indicate areas of upwelling mantle convection currents and may represent zones where lithosphere is being added to the Earth's surface. In this chapter we shall look at the most distinctive belts of high heat flow, ocean ridges and rift valleys, to examine the evidence that indicates how these features form and develop.

4.1 Ocean ridges

Linear ridge-like features can be found in most of the world's ocean basins (Fig. 1.1). The best explored, and perhaps the best developed topographically, is the Atlantic ridge. In some ways the word 'ridge' is a misnomer, since, in the case of the Atlantic, the feature is 2000 km or more wide and may occupy anything between one-half and one-third of the total ocean floor. At its highest points the ridge is fairly well defined and in plan faithfully follows the outline of the adjacent seaboards (Fig. 4.1). Reference back to Figure 1.1 will remind the reader that the Atlantic ridge is the only ocean ridge that can be considered 'mid-ocean'. At its margins the ridge grades into the abyssal plains of the ocean, with no obvious break (Fig. 4.2a); indeed, the distinction between abyssal plain and lower ridge appears to be academic. The abyssal plains have a remarkably smooth sediment-covered profile; but as the ridge is ascended, the seafloor becomes increasingly rugged, and subordinate ridges lying parallel to the ridge crest become common. The crest of the ridge itself is quite variable in height (Fig. 4.1), but over extensive sections it rises to less than 3000 m below the ocean surface. In some places the crest reaches the ocean surface, forming islands (e.g. Iceland, the Azores, Ascension Island, Tristan da Cunha).

If looked at in detail, the ridge crest often shows a

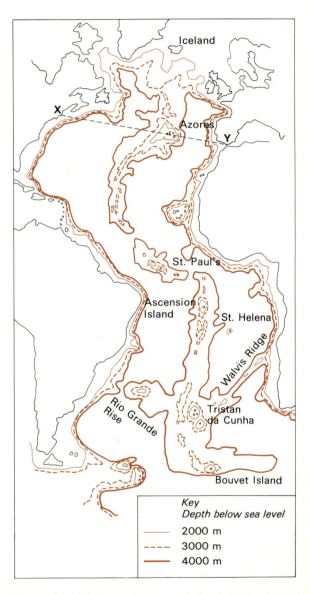

Figure 4.1 Bathymetric map of the Atlantic Ocean, showing the major submarine features.

central valley (Fig. 4.2b). Where detailed submarine mapping has been carried out, as in the part of the ridge lying southwest of the Azores (Fig. 4.3), the central valley turns out to be parallel to the main ridge, deep and straight. In the area of the map, the valley is 1600 m deep and 30 km wide. On its western side the valley slope is steep and straight and falls to a flat valley floor, whereas the eastern side of the valley is stepped in a series of terraces.

As Figure 4.3 suggests, the central valley continues in one direction for only a short distance before suddenly turning through 90°. The valley lying orthogonal to the ridge crest, however, is topographically less distinct, has little in the way of a flat floor and is not bounded by parallel valley sides. After another short distance another 90° turn is made, the valley resuming its normal course and topography. The form of the ridge-parallel sections of valley is so strikingly similar to continental rift valleys (see Section 4.2) that they have been so named. The central rift is obviously broken up into sections and offset by some kind of fault orthogonal to the central ridge. The presence of such **transform faults** is interpreted by submarine topography and seismic data in the first instance, but these structural elements have now been identified for the entire

length of the Atlantic ridge (Fig. 4.4). Beyond the ridges the line of a transform fault is usually continued into a more general **fracture zone**, which appears in the seafloor topography but is seismically less active.

Where there is only a thin veneer of sediment on the surface, the rocks from which the ridge is made can be seen. The dominant rock material is a volcanic lava of basalt composition (Appendix A.2). Because most of the volcanic eruptions along the ridge are submarine, the lava that escapes is chilled rapidly by contact with sea-water and solidifies into a mass of billowing protruberances known as **pillow lavas**. Typical pillow lavas are not quite true basalts, because they tend to contain a sodium-rich feldspar instead of the normal calcium-rich feldspar (Appendix A.1). Lavas of this composition are termed **spilites**. Whether the sodium replacement in the feldspar is due to chemical reaction with sea water is not clear, but they are common in submarine eruptions. Along the Atlantic ridge, spilite flows seem to occur quite often along the sides of the central rift valley.

Successive volcanic eruptions at a particular site may pile up on the ridge until the ridge breaks up through the ocean surface to form an island.

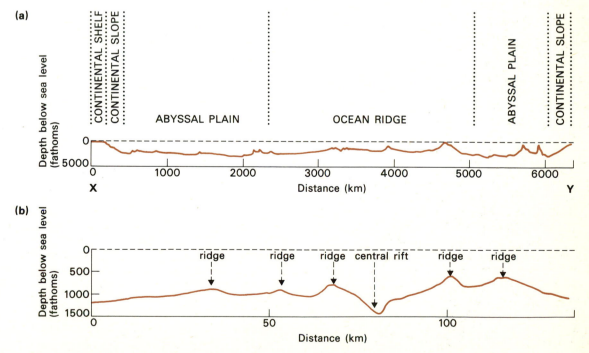

Figure 4.2 Cross-profiles of the Atlantic seabed: (a) along a line from New England to Gibraltar (shown as X–Y on Fig. 4.1); (b) in more detail, across the portion of the central ridge shown in Figure 4.3.

Geologically speaking, new islands of this sort are constantly being created along the line of the ridge. One of the most recent events was the appearance of the island of Surtsey off the southern coast of Iceland in 1962.

All the other islands that appear along the Atlantic ridge are of volcanic origin. Some (e.g. St Helena) seem to be dormant, whereas other appear to be highly active; the entire population of Tristan da Cunha had to be evacuated during the 1962 eruption, for example. By far the largest of these islands is Iceland itself, where the whole land area is underlain by basalt lavas. We shall be discussing volcanic eruptions on Iceland in more detail in Chapter 7, but for the moment it is the age of lavas on Iceland that is of paramount interest. Figure 4.5 shows that the most recent lavas (Holocene, i.e. later Quaternary) lie along a belt that broadly

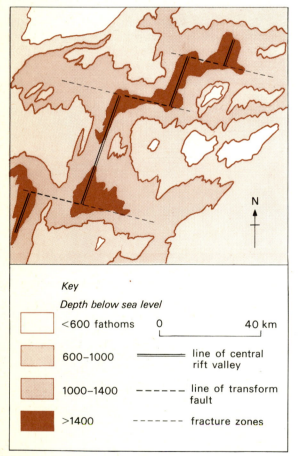

Key

Depth below sea level

☐	<600 fathoms	0	40 km
▨	600–1000	═══	line of central rift valley
▦	1000–1400	- - - - -	line of transform fault
■	>1400	- - - - - -	fracture zones

Figure 4.3 A detailed relief map of a small part of the Atlantic ridge southwest of the Azores, showing the central rift valley offset by transform faults (from Heirtzler & Bryan 1975).

passes through the centre of the island along a SW–NE axis. Apart from the fact that all the active fissures and vents in Iceland lie along this belt, the SW–NE axis appears to represent a section of the crest of the Atlantic ridge (Fig. 4.4) offset by transform faults. Away from this central axis the lavas become progressively older in either direction.

This same phenomenon is also seen in the basalt rocks of the submarine ridge but in a rather interesting manner. We have already noted that it is possible to 'read' the pattern of magnetism that existed when lavas were originally erupted (Section 2.1.5), because iron-rich minerals tend to become induced with a magnetism of their own. When magnetometer surveys are carried out across an ocean ridge, a very strange pattern emerges. As Figure 4.6 shows for a section of the Atlantic ridge southwest of Iceland, the basalt rocks appear to be showing magnetic striping. The coloured stripes show seafloor that has a strong magnetic field, but the blank areas show sections of the seafloor where the magnetic field is rather weak. The stripes lie parallel to the ridge crest and, if Figure 4.6 is examined closely, show a rather remarkable repetition of stripe pattern on the two sides of the crest. The key to the pattern is what is known as **polarity reversal**. In the chapter on continental drift it was argued that polar-wandering curves represent movement of the continents and not of the magnetic poles. Although it may remain true that the north magnetic pole, for example, does not wander at random over the Earth's surface, it appears that, periodically, the entire magnetic field of the Earth reverses, making the north magnetic pole the south and vice versa. As far as the seafloor is concerned, if we assume that the present geomagnetic field is 'normal', any basalts erupted during normal field conditions will have an induced magnetism that corresponds to the prevailing geomagnetic field and consequently will show up on a magnetometer survey as areas of strong magnetism. Any basalts erupted during periods of 'reversed' geomagnetic field, however, will contain an induced magnetism that lies in the opposite direction to the present-day geomagnetic field, and these areas will be revealed by magnetometer survey to have weak magnetism, since the induced magnetism of the rocks will 'work against' the present geomagnetic field.

Independent evidence from continental lavas has shown that polarity reversal does not operate on a regular basis but may occur with time intervals from a few thousand up to about a million years. By dating lavas in which polarity direction has been

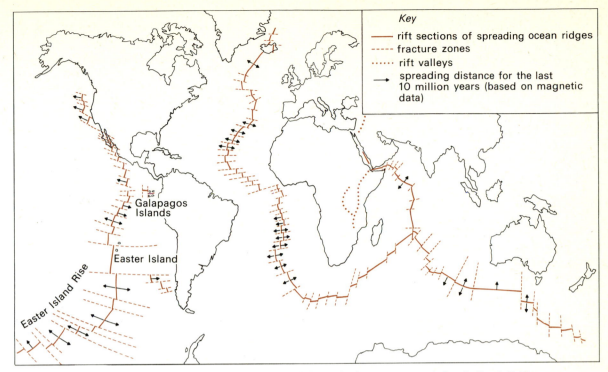

Figure 4.4 The world rift system, together with some of the main fracture zones (after Bullard 1969).

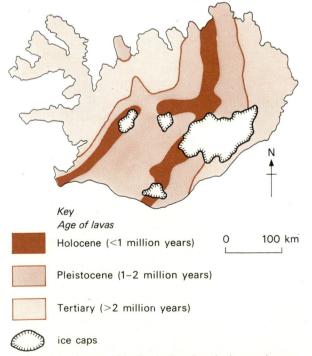

Key
Age of lavas

Holocene (<1 million years)

Pleistocene (1–2 million years)

Tertiary (>2 million years)

ice caps

0   100 km

Figure 4.5 A map of Iceland, showing the increasing age of lavas away from the central SW–NE axis (from Rice 1977).

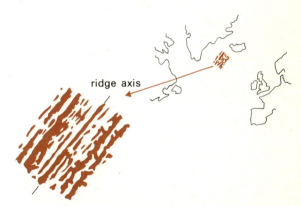

Figure 4.6 The pattern of magnetic striping in the basalt rocks of the oceanic crust on the Atlantic ridge southwest of Iceland. Solid colour shows zones within which a 'normal' magnetic field is induced in the lavas, and the blank areas show 'reversed' magnetism (after Heirtzler, Le Pichon & Baron 1966).

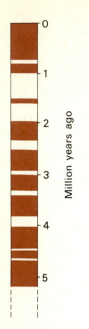

Figure 4.7 A time scale for periods of normal (solid colour) and reversed (white) geomagnetic polarity during the last 5 million years.

Million years ago

recorded, it is possible to construct a time scale for 'normal' and 'reversed' polarity (Fig. 4.7). Armed with this piece of information, we can now interpret the pattern of magnetic striping on the seafloor. First, it can be seen that the age of basalt increases away from the ridge crest. Secondly, precisely the same pattern is repeated on both sides of the central ridge.

At the present day, the only location for volcanic eruptions in the Atlantic is along the crest of the ridge. Assuming that this was the case in the past, we have some exceptionally good evidence in the magnetic-striping pattern for **seafloor spreading**. Basalt is erupted at the crest of the ridge, and over time the seafloor spreads away from the crest in opposite directions, leaving room for new basalt to be erupted. Since the magnetic pattern also allows dating of the basalts, it is possible to estimate the rate at which the seafloor is opening up. In the case of the North Atlantic, this works out at approximately 2 cm/yr, so that the oldest seafloor basalts, off the coasts of the United States and West Africa, are about 200 million years old. This figure

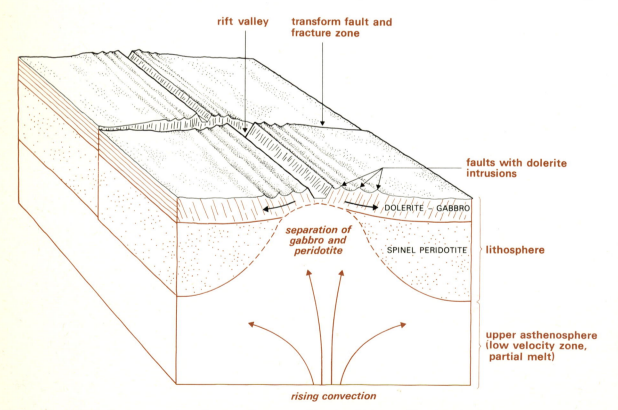

Figure 4.8 Schematic representation of a spreading ocean ridge, in which upwelling mantle convection currents initiate the ridge, which then develops by rifting and magma intrusion. New lithosphere is created by these processes.

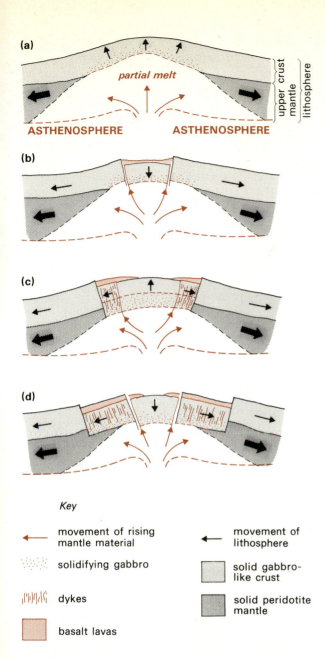

Figure 4.9 The development of an ocean ridge. (a) Upwelling convection in the mantle causes the oceanic crust to form a ridge. (b) Lateral tension develops, causing rift faulting and the downward movement of the central block. Magma intrudes along the faults, giving surface lavas. (c) Lateral movement continues, initiated by further intrusions parallel to the original rift faults. (d) The main rifting sequence is repeated periodically as upwelling continues.

establishes the 'age' of the Atlantic, and it corresponds very well with the independent evidence from continental drift.

What processes, then, are operating beneath an ocean ridge that are capable of causing seafloor spreading and producing the topography of the ridge? Figure 4.8 is a schematic model that allows one possible interpretation of the data. The moving force beneath the ridge is an upwelling convection current of mantle material, rising from at least the lower asthenosphere. We have already presented the heat flow data (Fig. 3.9) that indicates a rising convection current beneath the ridge, and further support is given by seismic studies, which suggest that the Mohorovičić discontinuity is not clearly defined beneath the ridge.

The effects of the upwelling can be illustrated in an evolutionary diagram like Figure 4.9, although the continuous nature of the processes should be emphasised. Were a rising current to start up beneath solid oceanic crust (Fig. 4.9a), the crust would be raised into a ridge. The upward thrust so created, combined with a tendency for the crust to fall away laterally under the influence of gravity, would produce faulting and eventually a central rift valley (Fig. 4.9b).

Meanwhile, as the mantle material ascends towards the surface, it seems to go through several changes of state. At some point around 250 km below the surface, the mantle material seems to melt into a semi-liquid state as pressure decreases. The lateral spreading of such semi-liquid rock seems to give rise to the low velocity zone of the upper asthenosphere. Above about 100 km below the surface, some of the denser minerals of the semi-liquid material spread laterally and solidify into the peridotite layer of the lower lithosphere, while lighter material rises higher and solidifies eventually into gabbro to form the base of the oceanic crust, which is constantly moving away from the ridge. As faulting occurs, lines of weakness are created in the upper crust, along which still-liquid material of gabbro composition can rise. This **magma** may reach the surface, in which case lavas are erupted, or it may cool within the fault lines to produce a **dyke**. Intrusion of magma along dykes may give added lateral force to the spreading process. Both lavas and dyke rocks will be of gabbro composition; but cooling more rapidly, they will produce rocks of finer texture, **basalt** and **dolerite** respectively (Appendix A.2). Consequently, as spreading from the ridge occurs, new oceanic crust material is created to fill the gaps so produced. A section through oceanic crust is likely to show lavas near the surface, dykes beneath and undifferentiated gabbro beneath that (Fig. 4.9c). Since peridotite is also separating out from the rising material between the crust and asthenosphere, the ocean ridge is actually producing the entire lithosphere layer.

Over time, the sequence of upwelling, faulting and igneous activity is repeated continuously. Earlier rift valleys disappear as lateral spreading occurs, pushing the rift slopes away to form subordinate parallel ridges, which become more and more degraded away from the crest and eventually form the floor of the oceanic abyssal plains (Fig. 4.9d).

The processes outlined above seem to occur directly beneath the central rift on the ocean ridge; but as we have already seen, the rift sections are fairly short, being constantly interrupted by fault lines lying at right angles to the line of the ridge. As Figure 4.4 indicates, the offsetting of the ridge is greatest where the changes of ridge direction are greatest. The transform faults exist, at least in part, to make such changes in direction possible. In Figure 4.10a a curving ridge is illustrated, and the arrows of seafloor spreading show that such an arrangement would cause the crust to converge at some points and diverge at others (leaving nasty holes in the surface!). Transform faults make equal seafloor spreading possible while the ridge is changing direction.

Figure 4.10b also makes clear the difference between the transform fault itself and its continuation in a fracture zone. The transform fault itself lies between two offset rifts, and consequently the relative motion of the seafloor on either side of the fault is in opposite directions. This opposing motion gives rise to most of the earthquakes along an ocean ridge. Beyond the central rifts, however, seafloor spreading out from adjacent sections of central rift is moving in a single direction, and the line of meeting is best described as a fracture zone rather than a fault, since there may be no relative movement on the two sides.

Although we have tended to concentrate upon the Atlantic in this section, the same topographic, geological and seismic data exist for most of the world's ocean ridges. It is, for example, possible to draw a world map showing the existence of spreading rifts and fracture zones (Fig. 4.4). As geomagnetic surveys become more complete, it is also possible to make comparative estimates of the rate of seafloor-spreading. As it turns out, the North Atlantic ridge, with its opening rate of 2 cm/yr, is rather slow in world terms. The South Atlantic

opens at 3 cm/yr, and the northeastern and southeastern Pacific rises give figures of 6 and 10 cm/yr respectively. An indication of relative spreading rates is given in Figure 4.4 by the arrows, whose length corresponds to the length of seafloor created in the last 10 million years on the basis of magnetic evidence.

A comparison of Figure 4.4 with Figure 3.12 will demonstrate that each of the plates defined by earthquake and volcanic distributions includes at least one margin where a spreading ridge is in the process of creating the lithospheric material of which the plate is made.

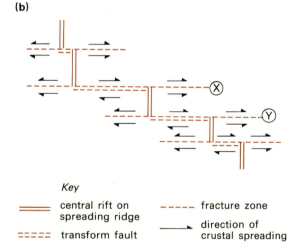

Figure 4.10 The role of transform faults. (a) Along a curved rift, spreading seafloor would converge at points marked A and diverge at points marked B. (b) Transform faults offset the rift, which follows the same plan, but spreading is now uniform away from the central ridge. Opposing arrows along transforms X and Y indicate opposite plate motion.

4.2 Continental rift valleys

Figure 4.4 shows that the major rift systems of the world occupy the crests of ocean ridges. The term 'rift valley', however, was originally applied to surface features in continental areas. A rift valley is a long narrow depression bounded by straight steep slopes, which originate from faulting of the Earth's surface. As such, rift valleys are the products of earth movements rather than denudation. Many examples of rift valleys, large and small, can be found on the continental areas of the world. Among the better known are the rifts that enclose Lake Baikal, in Russian Mongolia, and the River Rhine between Basel and Mainz. Closer to home is the Central Valley of Scotland (see Ch. 6.5), but this, like the Rhine and Baikal rifts, is fairly old, geologically speaking, and no longer represents an active tectonic zone. Two continental rift systems are, however, shown in Figure 4.4, as they are usually taken to be, if not currently active, at least recently active. These two systems are the Jordan Valley and, on a much larger scale, the East African rift system.

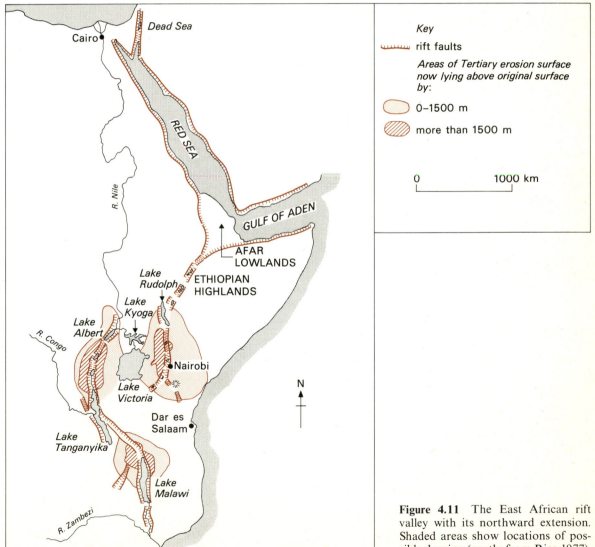

Figure 4.11 The East African rift valley with its northward extension. Shaded areas show locations of possible doming (partly from Rice 1977).

The East African rift system can be traced as far north as the Afar Lowlands (Fig. 4.11), from where a discontinuous valley crosses the Ethiopian Highlands to Lake Rudolf and then passes through a series of minor lakes to Lake Malawi and on to the Zambesi. A branch of the system passes west of Lake Victoria and includes Lakes Albert, Edward, Kivu and Tanganyika. Throughout the system, the valley is about 40–50 km wide, although the depth and floor level vary considerably. The sides are steep and seem to be fault guided. Periodically, the valley has been dammed by lava flows, and many extinct volcanoes are associated with the faults along the valley sides.

For many years, constant argument surrounded the origin of this rift system. Discussion centred on either a tensional or a compressional cause of the faulting (Fig. 4.12). Both possible origins have problems on the basis of the available evidence, and over the last few years opinion has swung towards a cause that emphasises vertical rather than horizontal movements. It seems quite likely that upwarping of the continental crust occurred along the line of the rift system during the early stage of development. Careful geomorphological analysis has allowed the interpretation shown in Figure 4.14. As this diagram suggests, an early westward-flowing drainage system, which had created a low-relief denudation surface (stage 1), was progressively interrupted by upwarping and subsequent rifting, with the diversion of drainage into the opening rift valleys (stage 2). The rate of upwarping eventually

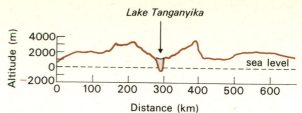

Figure 4.13   Cross-profile of the western rift of the East African rift system, on an east–west section through Lake Tanganyika.

exceeded the downcutting power of some of the westward-flowing rivers, which backed up to form the shallow Lakes Victoria and Kyoga (stage 3). The extent of upwarping can be seen today both in the topography of the rift valley sides (Fig. 4.13) and in the deformation of the denudation surface produced at stage 1 (Fig. 4.11).

The emphasis upon the importance of upwarping along the rift system makes it very tempting to draw a direct parallel with the processes at work beneath an ocean ridge. Does the East African rift system represent convection currents within the mantle rising beneath a section of continental rather than oceanic crust? Is Africa splitting apart along the line of the rift? A nice idea though this is, the evidence, unfortunately, does not support it. The main difficulty lies in the age of the rift system. Although the rifts are very narrow and show no sign of revealing rock lying beneath the continental crust, the associated earth movements may, in some places, have started as much as 200 million years ago; in other words, little progress has been made in opening the continent up. Similarly, although volcanoes are common throughout the system, most are extinct; and despite the presence of the rift faults, the level of earthquake activity is lower than may be found along an ocean ridge. Together, these facts indicate that the East African rift system is no longer very active from a tectonic point of view and can hardly represent current continent-splitting.

4.3   The breakup of continents

Although the East African rift system does not seem to represent current continental breakup, there are two locations in the world (Fig. 4.4) where an oceanic rift system seems to disappear beneath a continent: the Gulf of California and the Red Sea. Detailed submarine surveys of the Gulf of California have shown (Fig. 4.15a) that the floor of the Gulf consists of short sections of a rift system, offset by extensive transform faults. The northern exten-

(a)

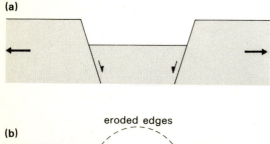

(b)

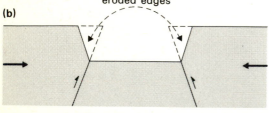

Figure 4.12   The origin of rift valleys. (a) Crustal tension causes a central 'keystone' block to fall downwards. (b) Crustal compression forces a central block to be pushed under.

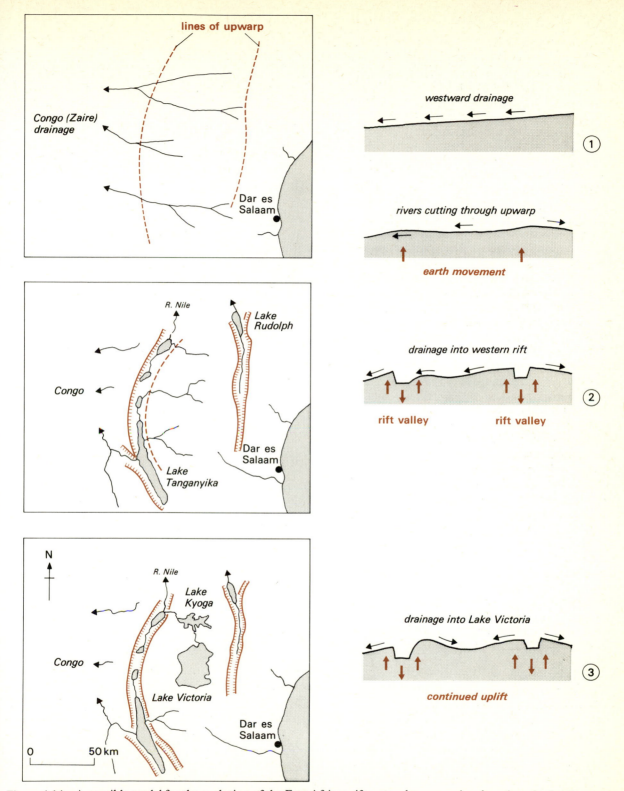

Figure 4.14 A possible model for the evolution of the East African rift system by upwarping, based on the development of the drainage pattern (see text for details) (partly from Rice 1977).

sion of these transform faults is the San Andreas Fault system, which cuts through continental crust. The central rift shows high heat-flow values, and there is a considerable amount of additional evidence to show that seafloor spreading away from the central rift is causing the peninsula of Baja California, and its northern extension into coastal California, to move northwards along the Pacific coast relative to the rest of the continent.

The Red Sea lies between the clear rift valley of the Jordan to the north and the Afar Lowlands to the south, which mark the extremity of the East African rift system. The steep slopes along the coast of the Red Sea are often drawn as a rift valley (Fig. 4.11), making a continuous system running from the Jordan to the Zambesi. To the east, however, the Red Sea connects with the Indian Ocean through the Gulf of Aden. The clearly defined spreading ridge of the Indian Ocean with its offset rift can be clearly traced into the Gulf of Aden. Within the Red Sea, however, the situation is slightly different. Most of the floor of the Red Sea is fairly shallow and consists of sediments overlying continental crust (i.e. granite and grandiorite-type rocks). Along the centre, however, is a deep central rift (Fig. 4.15b) cutting through continental crust, at the bottom of which are basalt-type rocks (i.e. oceanic material). The bottom of this deep trench yields high heat-flow values, and typical magnetic-striping patterns can be seen over short distances, which not only confirm seafloor spreading but suggest that the central rift is not more than about 10 million years old. The obvious implication is that seafloor spreading beneath the Red Sea is causing the overlying continental crust to split and move apart. Eventually, Arabia may be split off from Africa.

During investigations of this kind of situation, one particular feature seems to have sparked off a good deal of interest. Some writers have suggested that the Gulf of Aden, the Red Sea and the Afar Lowlands represent the three rifts radiating out from a single point of uplift. Although a single point of uplift is not clearly defined in this instance, there is some evidence to suggest that the area was originally domed over a point close to where the three rifts meet. There is also a suggestion that the uplifting forces that produced the East African rift system were domes rather than lines of upwarp (Fig. 4.11). These features have been used as evidence for the operation of **hot spots** within the Earth's mantle.

So far we have only considered convection currents within the mantle as operating along lines. The 'hot spot' theory suggests that rising plumes of

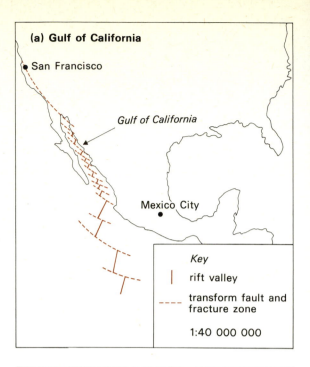

(a) Gulf of California

San Francisco

Gulf of California

Mexico City

Key

| rift valley

---- transform fault and fracture zone

1:40 000 000

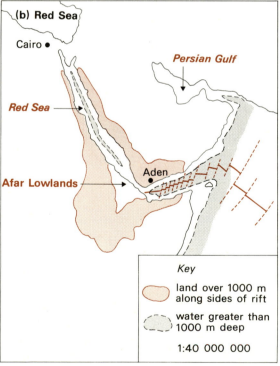

(b) Red Sea

Cairo ●

Persian Gulf

Red Sea

Afar Lowlands

Aden

Key

land over 1000 m along sides of rift

water greater than 1000 m deep

1:40 000 000

Figure 4.15 Two examples of ocean ridge systems disappearing beneath continental crust.

mantle material occur at isolated points on the surface. A hot spot beneath oceanic crust may produce only a volcano (see next chapter), but beneath a continent the rising plume of material would cause the crust to dome. Experiments have shown that, if brittle material is forced into a dome shape, it tends to crack, giving radial fractures (Fig. 4.16). Most commonly, three more or less equally spaced radial fractures appear, which in the case of a continent would evolve into rift valleys.

Figure 4.17 shows the theory used in an attempt to explain the shape of the Atlantic split. If hot spots were located as shown, the direction of the radial fractures would tend to be controlled by the distance to the next hot spot; that is, fractures on

adjacent domes would tend to align. Continental splitting would occur along the line of maximum stress and would therefore follow the line of shortest paths between hot spots. In some cases all three arms of a dome would fracture fully, giving rifts extending right through the continental crust. A radial fracture pointing straight into a continent and away from any other doming points, however, might never open fully and would form a failed rift. Observers claim that failed rifts, represented now by narrow, deep but sediment-filled troughs called **aulacogens**, can be found at various points along the Atlantic seaboards. The Niger and Amazon are said partly to occupy aulacogens.

The hot spot theory of continent-splitting is, it must be emphasised, a theory that has not yet found general acceptance. It may be capable, however, of explaining the overall shape of continents as well as some ocean floor features. The North Sea, for example, is partly underlain by a rift valley (Fig.

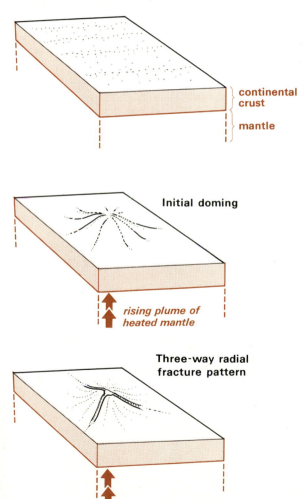

continental crust

mantle

Initial doming

rising plume of heated mantle

Three-way radial fracture pattern

Figure 4.16 Diagrammatic representation of the doming and radial fracture of continental crust by an upwelling 'hot spot'.

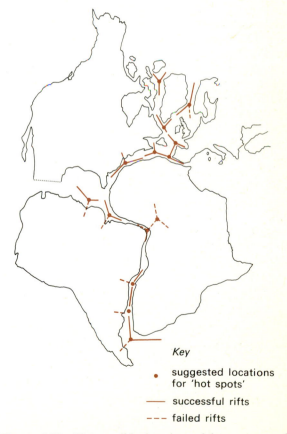

Key

• suggested locations for 'hot spots'

— successful rifts

--- failed rifts

Figure 4.17 The possible location of hot spots and associated rifts involved in the opening up of the Atlantic (after Burke & Wilson 1976).

4.17), which may represent a failed arm. The distinctive diagonal ridges connecting the ocean ridge in the South Atlantic with the opposing seaboards (the Rio Grande and Walvis Ridges, Fig. 4.1) may represent accumulations of lavas carried away from a particularly active hot spot by seafloor-spreading. In broader terms, the theory raises the interesting question as to whether the upwelling mantle convection currents that exist along the ocean ridges between potential hot spots are the *result* of surface-fracturing guided by a hot spot pattern.

Quite what position the East African rift system occupies in this discussion is unclear. The rift valleys may represent continental splitting that never really got going. Nonetheless, there is little doubt that continental splitting in the past broke Pangaea up into its constituent parts, and similar processes operate today in places like the Red Sea.

Once a continent has been split apart, there is relatively little evidence left on the edge of the continental crust to indicate that seafloor spreading was the agency responsible, unless it is the relics of a domed rift, such as exist along the edge of the Red Sea. The presence of large spreads of basalt lavas, however, may be a consequence of seafloor spreading. As already indicated, continents are underlain primarily by igneous rocks of granite or granodiorite composition, and most continental volcanoes tend to be of at least similar chemical composition (andesite or rhyolite, Appendix A.2). Ocean ridges, however, are typically made of basalt lava, because the magma originates within the Earth's mantle. We have already seen that islands such as Iceland are made of numerous super-imposed basalt flows. If a continent splits above an incipient ocean ridge, basalt lavas may pour out on to the continental surface. Basalt flows are associated with the East African rift valley, and closer to home they can also be found in northern Ireland and western Scotland (Fig. 6.24). The British Isles now lie some distance from the North Atlantic ridge (Fig. 4.1); but as Figure 4.17 shows, the splitting of the North Atlantic must have been preceded by rifting at no great distance from northwestern Scotland. The evidence from seafloor magnetism indicates that the North Atlantic has been opening progressively over time from the south (see also Fig. 2.6), so that, whereas Florida and the Gambian coast may have rifted apart 200 million years ago, the rift separating the British Isles from Greenland was probably active within the last 50 million years. The basalt plateaux of Antrim and the Inner Hebrides are of comparable age with that rifting episode.

Other areas of the world have basalt plateaux on a much larger scale than this and over a range of geological periods. These **flood basalts**, as they are sometimes known, poured out across the Parana Basin of southern Brazil in Jurassic times, the Deccan Plateau of India during the Eocene (the same as the Antrim basalts) and the Columbia–Snake River Plateau of the northwestern United States during the last 20 million years. The Parana and Deccan basalts can easily be referred to continental splitting over a spreading ridge. The Columbia–Snake River setting is rather more complex and indicates the mixed history of the western coast of North America.

Chapter 5

The Surface of the Moving Plates

At the constructive plate margins described in the last chapter, upwelling mantle material creates new lithosphere. As a result oceanic crust spreads away from the ocean ridge system to form the surface of a moving Earth plate (Fig. 3.12). Where spreading ridges occur beneath continents, the continental crust is split apart and is eventually carried away from the ridge by the outward-spreading oceanic crust. In this chapter we are going to look at the oceanic and continental areas of the Earth plates as they move away from the constructive margins, in order to see what changes occur at the surface before they come within the active zones to be discussed in Chapter 6.

5.1 Volcanic activity on the plate surface

Since the surface extent of the lithospheric plates is partly defined by the active volcanic and earthquake belts, it is to be expected that volcanoes are somewhat rare beyond the active margins. Nonetheless, as Figure 3.10 indicated, volcanoes do occur at a few isolated spots outside the main belts.

The Pacific Ocean (Fig. 5.1) largely consists of a single lithospheric plate. A discontinuous spreading ridge runs along the eastern margin of this unit, and magnetic surveys of the seafloor indicate that the oceanic crust of the Pacific is moving broadly to the northwest away from the Easter Island Rise. The northern and western margins of the Pacific are marked by the island arcs and deep ocean trenches of another active zone. On the plate surface itself are many islands, most of them forming linear groupings and all consisting of a volcanic core, even if that is no longer visible at sea level. Nonetheless, away from the active margins the only volcanoes actually active today are those found on the Hawaiian Islands.

Hawaii itself is simply one island at the end of a submarine ridge stretching some 3000 km to Midway Island. Since the ridge and all the islands along it are dominantly composed of basalt lavas, it is tempting to assume that this is another example of an oceanic spreading ridge. However, the ridge does not show any sign of a central rift, nor is there any seafloor magnetic evidence for spreading away from the ridge. Furthermore, whereas active volcanoes may be found all along a spreading ridge, in the Hawaiian group the only active vents are on Hawaii itself. Interestingly, dating of lavas on the other islands shows that only one point on the ridge was active at any one time. The active centre, moreover, seems to have moved to the southeast over time (Fig. 5.2).

All this points to a magma source rising up over a small area rather than along a line. That, of course, sounds remarkably like the 'hot spots' introduced in the last chapter. One consequence of the hot spot theory is that the plumes of rising magma forming the 'spots' originate at much greater depth in the mantle than the more generalised lines of upwelling under a normal spreading ridge (a reasonable supposition, of course, if the spreading ridges are simply caused by the fracture zones between hot spots). Chemical analysis of the Hawaiian basalts tends to confirm that this is in fact the case.

Is the whole ridge, then, the product of a hot spot moving progressively to the southeast? Figure 5.1 makes reference to two other Pacific island chains: the Pitcairn–Tuamotu and Austral groups, where a linear ridge of basalt also shows lavas decreasing in age to the southeast. Obviously, it is possible to envisage three separate hot spots all moving in the same direction, but it is more likely that the three hot spots were stationary while the Pacific plate moved as a single unit to the northwest. Such a

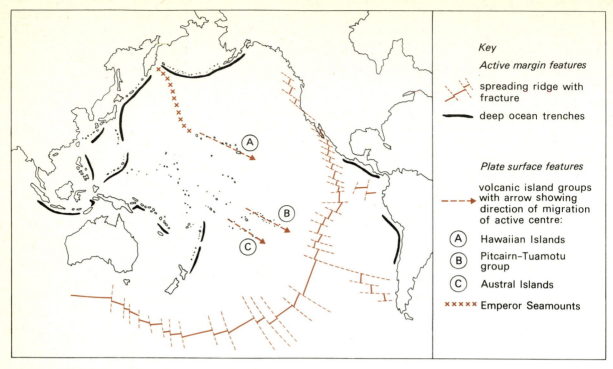

Figure 5.1   The Pacific plate, with its tectonic margins, showing island groups of basalt volcanoes.

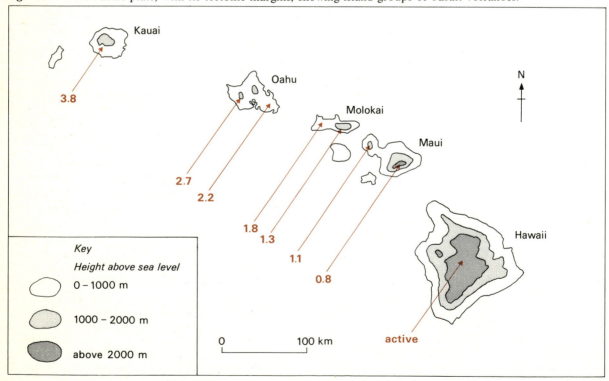

Figure 5.2   The Hawaiian Islands. Figures in colour indicate the age of the youngest basalt lavas on each island in millions of years.

conclusion would be consistent with the direction of seafloor spreading from the Easter Island Rise, and it would also confirm that the floor of the Pacific is moving as a single unit. The Emperor Seamounts, lying between Midway Island and the Aleutians (Fig. 5.1), seem to mark an even older extension of the Hawaiian ridge, with lavas varying from 25 million years old at Midway to 75 million years old at the Aleutians. The line of this extension suggests that the direction of Pacific plate movement has changed over time.

Many other island groups lying on the surface of the Pacific plate (Fig. 5.1) probably have a similar origin. In all cases the core of the island is volcanic and of basalt composition, although the lavas are frequently hidden below sea level or later coral growth.

Away from the Pacific plate active volcanoes outside the main belts are virtually unknown, which again suggests that hot spots *create* spreading ridges and are not normally found isolated on a plate surface. It is, incidentally, worth remarking that the hot spot model of Atlantic ridge evolution (Section 4.3) includes the production of basalt ridges (e.g. the Walvis and Rio Grande Ridges) on the plate surface as it moves away from the hot spot. Volcanoes on the African plate are similarly associated with hot spots and, in this case, continental splitting and can therefore be regarded as lying on an active margin, even if the rift system is not actively developing today.

5.2 Denudation and sedimentation

The lithosphere of the Earth is created at spreading ridges by processes powered by the internal energy of the Earth, which is in turn the result of the fission of radio-isotopes buried deep within the Earth. Having been created, the surface of the lithosphere is subsequently modified on its journey across the globe by the external processes of **denudation** and **sedimentation**. These processes are ultimately powered by solar radiation, which is responsible for the characteristics of climate as well as ocean currents and waves, and by gravity.

Denudation is the term applied to the overall breakdown of the landscape by external processes and is usually thought to consist of **weathering**, which is the *in situ* breakdown of rock by chemical or mechanical means, **transportation** of the weathered material by a moving agent (e.g. running water,

moving ice, wind or waves) and **erosion** of the underlying surface by the transporting agent. The transported material is frequently **deposited** for short or long periods of time when the energy of the transporting agent falls. If this material is trapped in a suitable environment for long enough, it may undergo the compaction and cementation that turn sediments into **sedimentary rocks** (Appendices A.3 and A.4).

Taking the world as a whole (Fig. 5.3), the areas of high surface relief, and therefore high transporting energy, tend to be dominated by denudation (although there are obviously local pockets of deposition). In contrast, sedimentation occurs in areas of low surface energy. A comparison with some of the earlier maps showing plate boundaries reveals that most active plate margins are areas where denudation dominates. Because they are tectonically and volcanically active, however, the internal forces are greater than the external processes of denudation, and the landscapes remain largely the products of those internal forces. Only when we move on to the plate surfaces away from the active margins is it possible to see landscapes where denudation is largely responsible for the surface features.

In tectonically stable areas (i.e. the plate surfaces away from the active margins), where denudation dominates, the landscape can be thought of as the product of three factors:

(a) *The nature of the denudation processes.* Major denudation systems are controlled largely by climate; thus, there are humid systems where rivers are the main transporting agent, glacial systems where moving ice is important, and deserts where, in the absence of other agents, wind often becomes significant. Each of these systems tends to produce distinctive landscape features, and the reader is directed for more detail to a geomorphology text such as *Landscape processes*.

(b) *The nature of the underlying bedrock.* The resistance of rock to breakdown and the structure of the rock (folding, faulting, etc.) continue to exercise an important role in the overall shape of the landscape, even when the structures are not actively being created.

(c) *Time.* This factor should also be regarded, for two reasons:
   (i) The longer a particular set of denudation

processes operates upon an area, the greater the total amount of surface removal. Most geomorphologists envisage a **cycle of denudation** in which a landscape is progressively levelled over time to produce an area of low relief, often referred to as an **erosion surface** but perhaps more correctly described as a **planation surface**. Strangely enough, it is very difficult to find many extensive planation surfaces actually in the process of forming today. We shall return to this point in Section 5.4.

(ii) The nature of the denudation system affecting a particular area may change over time. This may be due to changes in the world climate, particularly those caused by glacial epochs, or it may be due to changes in the relative levels of land and sea, which bring marine erosion inland or expose submarine areas to subaerial denudation.

Over most of the tectonically stable areas of the world plates, both continental and oceanic, sedim-

entation is dominant (Fig. 5.3). Sediments tend to collect in distinctive **sedimentary environments**, which have marked topographic features and which, when present on a large scale, form important features on the Earth's surface. Although we cannot possibly examine the entire range of sedimentary environments to be found on the Earth's surface, it is important to be familiar with at least the major types.

5.2.1 Glacial environments

Present-day glacial deposition is naturally largely impossible to observe, since it is taking place beneath the major ice sheets (e.g. Antarctica, Greenland) and only the marginal areas are visible. Nevertheless, the world has experienced repeated glacial phases during the last few million years, with the result that very extensive areas of the northern hemisphere have been covered by ice sheets in the recent past. Areas which were close to the centre of ice dispersal (e.g. the Baltic, Labrador) tend to have a rather thin veneer of glacial debris, being dominantly zones of erosion. Glacial debris, in the form of

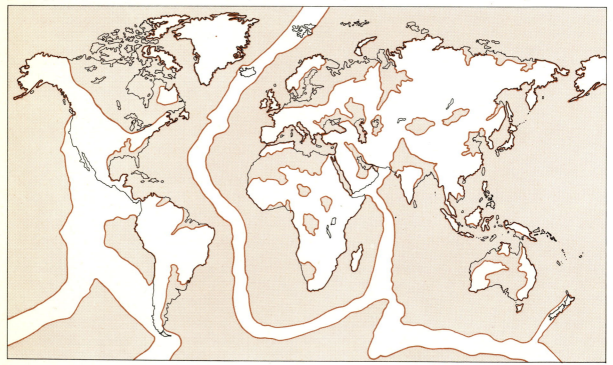

Key

◯ areas dominated by active, or recent, denudation

◯ areas dominated by active, or recent, deposition

Figure 5.3 The dominant areas of denudation and deposition in the world in recent times (from Allen 1975).

poorly sorted **moraine**, increases towards the edge of the ice. Much of this moraine simply forms a blanket of undifferentiated **till**, filling pre-existing valleys and leaving a low relief surface. Streamlined moraine ridges (e.g. **drumlins**) occur within the zone of active ice movement, but these give way to **terminal moraines**, lying parallel to the ice front, as the edge of the ice sheet is approached. Today, extensive terminal moraines can be found in the area south of the Great Lakes in the USA and on the North German Plain in Europe.

5.2.2 Deserts

Contrary to popular imagination, many deserts include river systems, although they are typically intermittent (i.e. active only periodically) and flow into basins of internal drainage, because rates of evaporation are generally high compared with available precipitation. Where rivers flow from mountain areas into desert basins, **alluvial fans**, consisting of cone-shaped accumulations of coarse material, occur. Towards the centre of desert basins, wide plains of fine silt and clay laid down by intermittent rivers give way to **playa lakes**, which may evaporate after rainfall to leave salt flats. These general relationships can be seen in the arid zone of Australia, where alluvial plains run westward from the Great Dividing Range toward the clay plains and playas of the Lake Eyre basin.

The most distinctive feature of the desert environment, however, is the accumulation of wind-blown sand in major sand seas, called **ergs**, such as the Gibson Desert of Australia. Ergs can be found in most desert areas of the world and many include very extensive dune systems whose lineations reflect the pattern of prevailing winds.

5.2.3 Fluvial environments

The river is the most important transporting agent over much of the world's land surface. Apart from localised accumulations of weathered material on hillslopes, the main fluvial sedimentary environment is the river floodplain. This usually consists of a long but narrow accumulation of layered sediments. Most rivers with fairly constant flow have a **meandering** pattern, in which the channel itself is often contained within natural banks of coarse material called **levées**, beyond which finer sediments are deposited on the backswamp of the floodplain during periods of high river flow. These features are particularly well developed along the lower course of the Mississippi. Rivers experiencing seasonal flow (e.g. at the exit of glaciers) often show a braided channel pattern and consist of rather coarser sediments. Many rivers flowing into the Arctic Ocean, such as the Ob, show these characteristics. Major **alluvial plains** are formed by rivers such

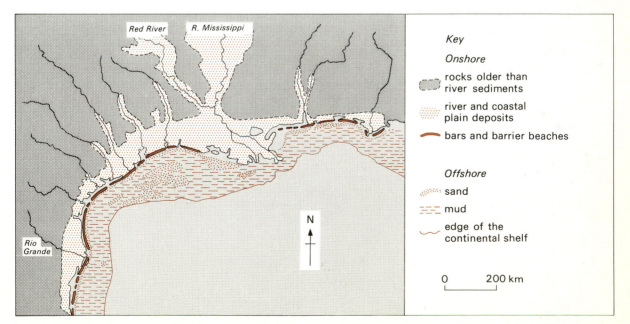

Figure 5.4 The main areas of coastal and continental shelf deposition in the Gulf of Mexico (after Allen).

as the Amazon, Mississippi, Tigris–Euphrates and Ganges.

5.2.4   Deltas
Where a river system brings more sediment into the sea than can be reworked by marine currents and waves, a delta may form. Deltas form major features in a number of places around the world. The shape of deltas may vary considerably. Where reworking by the sea is minimal, individual distributary channels push sediment lobes out to give the classic **bird's foot** shape found in the Mississippi delta. The Nile delta, however, has a smooth **arcuate** plan, formed by the sea creating a coastal barrier beach. The Ganges delta, meanwhile, has a coastline of mangrove swamp, encouraged by a high tidal range. All three deltas do show common characteristics as well. In particular, all have a flat platform crossed by distributary channels depositing coarse sand, between which are areas of fine mud accumulation. Offshore, the seaward side of the delta is a fairly steep slope of coarse material, grading into fine sediments beyond the delta slope.

5.2.5   Coasts
In many places the typical coastal deposit is the **bayhead beach**, which forms in any embayment between eroding headlands. In areas where coastal deposition is dominant, however, the main features tend to be a **barrier beach** or **offshore bar** parallel to the coast, behind which are shallow **lagoons**, which may fill with fine sediment to form a **coastal plain**. All these features are recognisable along the coastline of the Gulf of Mexico (Fig. 5.4). On a world scale, major coastal-plain accumulations tend to occur in tectonically stable areas, where either the prevailing wind is offshore (e.g. Gulf of Guinea, west Australia, southern Brazil) or where the land configuration protects from the prevailing wind (e.g. the Baltic, the southeastern Mediterranean).

5.2.6   Continental shelves
Beyond the tidal zone sediments may settle out on the continental shelves. Generally, the size of sediment tends to decrease away from the shoreline (Fig. 5.4), although this pattern may be complicated by rivers bringing coarse material into the sea, by redistribution in submarine currents or simply by relics of former beaches, now submerged, as in the Gulf of Mexico.

In areas of tropical continental shelves, where water circulation is somewhat restricted and evaporation rather high, deposition may be in the form of carbonate material. Carbonate minerals, like the calcite that forms limestone, are easily dissolved during weathering and enter the sea in solution rather than as particles. Dissolved carbonates may be removed from sea water to form the shells of marine invertebrates, but in most shelf areas the organic content amounts to only a small proportion of the total sediment. In areas such as the Gulf of Arabia, carbonates may precipitate directly from the sea to give carbonate sands and muds.

5.2.7   Abyssal plains
Although the abyssal plains cover much of the oceanic area, their importance as sedimentary environments is largely confined to a fairly narrow belt close to the continental shelf. Within this belt two kinds of deposition are important. First, fine material from the continents, suspended in ocean currents, is dropped on the ocean floor gradually to accumulate as mud. Secondly, **turbidite** deposits are produced by **density currents**, which are parcels of sediment-charged water moving along the seafloor under gravity because they are denser than the surrounding water. Many density, or turbidity, currents start on the steep continental slopes, where fine sediments are delicately balanced. Any sudden movement (e.g. an earthquake) may cause part of the slope to collapse and mix sediment with water. The turbidity current then rolls down the slope and out on to the abyssal plain. In some places continental shelves are dissected by **submarine canyons** (Fig. 5.5), which are the submarine continuations of major river systems. Close observation has suggested that river sediments entering the sea may help to initiate turbidity currents, which on the one hand erode the canyons and on the other carry coarse material right out on to the abyssal plains, where it is deposited as a fan.

Away from the maximum reach of turbidity currents, the amount of sedimentation on the abyssal plains decreases rapidly. In the centre of many ocean basins, in fact, there is virtually no sediment derived directly from the land and carried through the water. Instead, the deposits consist either of thin clays, which seem to come from wind-blown dust on the leeward side of the deserts, or **oozes** produced by various forms of surface plankton.

Taking the world as a whole, then, the surface of the Earth, formed at tectonically active zones, is modified in stable areas by the combined effects of denudation and sedimentation. Denudation only dominates a landscape after some period of stability. Sedimentation may rapidly cloak the surface with a thin veneer, but this is rarely very substantial in continental areas. It is only at the edges of continents, in coastal belts, in deltas, on continental shelves and at the edges of the abyssal plains that sediment may reach substantial thicknesses.

5.3 Eustatic and isostatic changes of land and sea level

In many parts of the world, an examination of the coastline will reveal features such as raised beaches or drowned valleys, which are a clear indication that the relative level of land and sea is by no means constant. In many cases both features can be seen at the same place, indicating a quite complex history of coastal formation.

Such changes in the relative level of land and sea are of particular interest to us at this point, for two reasons:

(a) Sea level changes affect the rate and form of denudation and sedimentation.
(b) It seems fairly clear that denudation and sedimentation *cause* changes in sea level.

For the purposes of this discussion, it will be helpful to distinguish between sea level changes produced by a change in the total volume of water in the oceans (**eustatic** changes) and changes produced by vertical movements of the land (**isostatic** changes).

5.3.1 Eustatic changes
Changes in the total volume of water in the oceans can only occur if that water is stored somewhere else. The oceans currently hold about 97 per cent of all water at the Earth's surface, leaving very little, in

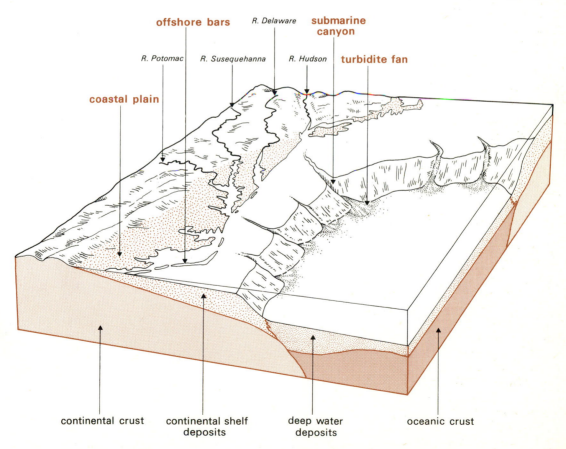

Figure 5.5 Block diagram to illustrate the main deposits of the continental shelf and abyssal plain off the eastern United States.

global terms, to be distributed between the atmosphere, lakes, rivers, soil, rock and ice sheets. However, of these, the ice sheets are by far the most important, storing about 75 per cent of all fresh water. Any change in the size of the ice sheets will cause worldwide sea levels to change. During the last 2 million years or so the world has been in an ice age, punctuated by numerous glacial periods, when ice sheets grew and sea level fell, as well as warm interglacials, when sea level rose. Figure 5.6 is an interpretation of data for the Mediterranean area, showing some of the eustatic changes that have taken place within the last 350 000 years.

The important points to emphasise about such eustatic changes are that they may occur very rapidly in geological terms and that they are global in their effects; Figure 5.6 should in theory be applicable to almost anywhere in the world.

5.3.2 Isostatic changes

The eustatic change graph (Fig. 5.6) shows that world sea levels have been rising for the last 10 000 years plus. A graph of sea level change for Norway over this recent period (Fig. 5.7), however, suggests the opposite; sea level seems to be dropping on the Norwegian coast. This paradox is usually explained by the theory of **isostasy**.

We have already seen (Ch. 2) that the Earth's crust consists of granite-like material (average density 2.7 g/cm³) overlying the mantle, which is made of denser (3.4 g/cm³) peridotite. In one sense this can be likened to the crust 'floating' on the mantle, because the mantle material does seem capable of some slow deformation. If the 'floating' analogy is maintained, it will be seen that the buoyancy of the

crust depends upon the displacement of mantle material below the crust. Consider, for a moment, an iceberg. Ice has only about nine-tenths the density of water, and an iceberg floats because the weight of water displaced by the ice below sea level is equal to the total weight of the iceberg – and everybody knows that only one-tenth of the iceberg actually protrudes through the surface. Figure 5.8a shows that the crust of the Earth is similar. Being about four-fifths the density of mantle material, it 'floats' with approximately four-fifths below and one-fifth above the equilibrium line (equal to sea level in the iceberg example). This isostatic balance explains why the thickness of the crust is much greater beneath mountains, since for every extra 1 km of height above sea level there must be an additional 4 km of crust below sea level to maintain equilibrium.

This is not a completely fixed balance, however, since any changes at the Earth's surface are obviously going to affect the total weight of the crust and therefore its equilibrium level. Denudation of the surface, for example, will remove weight and should cause isostatic rise, whereas sedimentation should cause isostatic fall. In practice, it is not quite that simple, because any point on the surface of the crust is partly held in position by the friction of adjacent pieces of crust (which is just as well, if you think about it, since otherwise life would be lived on a natural trampoline). Friction seems capable of holding the surface against minor movements. If denudation or sedimentation continues for long enough, however, the buoyancy forces will overcome frictional restraint, and vertical movements of the crust will result (Fig. 5.8). Vertical movement is not likely to be uniform over any great area, and so the gross effect is often a surface warping.

Isostatic changes have a number of interesting effects. Among the most pronounced are those connected with glacial episodes. The growth of ice sheets not only causes a worldwide eustatic drop in

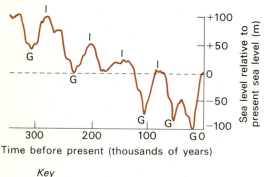

Figure 5.6 Possible eustatic changes in sea level during the Pleistocene for the Mediterranean area (after Fairbridge 1961).

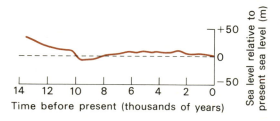

Figure 5.7 Changes in relative sea level during the last 14 000 years for the northern Norway coast (after Donner 1970).

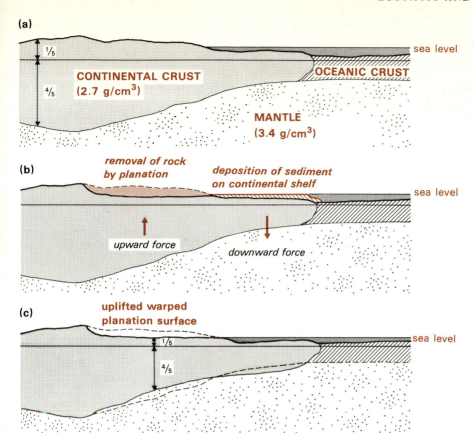

Figure 5.8 Diagrammatic illustration of isostatic equilibrium (see text for explanation)

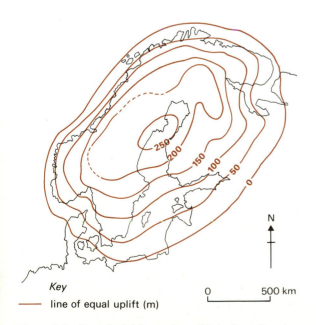

Key
— line of equal uplift (m)

0   500 km

Figure 5.9 Postglacial isostatic uplift in Scandinavia.

sea level; it also causes a more localised isostatic depression of the crust immediately beneath the growing ice (and a corresponding upwarp somewhere beyond the ice margin). Although this downwarping is not very easy to see in the world today, the results of the *last* glaciation are still having an effect. Careful surveying in Scandinavia (Fig. 5.9), for example, has revealed that the whole area is gradually warping upwards. What we have here is an example of isostatic readjustment of the crust following the removal of the weight of the last ice sheet. This explains the pattern of sea level change noted in Figure 5.7. The important point, of course, is that this is still happening some 10 000 years after the disappearance of the ice. In areas actually subjected to glaciation, of course, the pattern of sea level change during an ice age is enormously complicated, involving rapid eustatic changes at the beginning and end of each glacial, with overlapping periods of isostatic change. Figure 5.10 is an attempt to illustrate how these changes may appear for one particular area.

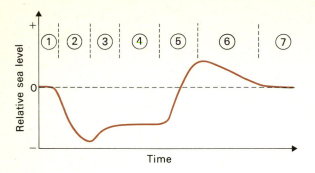

Figure 5.10 Hypothetical changes in relative sea level during a glacial: (1) preglacial sea level; (2) eustatic drop in sea level with ice sheet growth; (3) relative rise in sea level due to isostatic drop in land under the weight of ice; (4) steady glacial low sea level; (5) eustatic rise in sea level during melting of ice; (6) apparent drop in sea level due to isostatic readjustment of land as the weight of ice is removed; (7) postglacial sea level.

Under normal conditions, of course, isostatic changes accompany denudation and sedimentation. The tendency for the crust to rise isostatically is probably a major reason for the rarity of planation surfaces across the world's land surfaces, since in many instances isostatic uplift occurs before planation is complete. When this happens, the whole landscape is rejuvenated, and the rates of denudation are increased. In Africa a whole set of isostatically uplifted planation surfaces is still visible. In the extreme south, the summit plateau of some 3000 m in the Drakensbergs probably marks a planation surface of Cretaceous age. Further north, much of central Tanzania is dominated by a Miocene planation surface now standing between 1000 and 2000 m. In each case the surface appears to have formed close to sea level by escarpment retreat, with isostatic uplift occurring once sufficient material had been removed. The edge of the uplifted surface is often marked by a steep escarpment, and these escarpments form the sites of such natural spectacles as the Victoria Falls on the Zambesi, where the river drops from a higher to a lower planation surface.

Sedimentation produces isostatic downwarping. Local effects can be seen around deltas, for example, but the process is more effective on the edge of continental shelves and the adjacent abyssal plains. In both cases downwarping allows further sedimentation to occur, and this is particularly

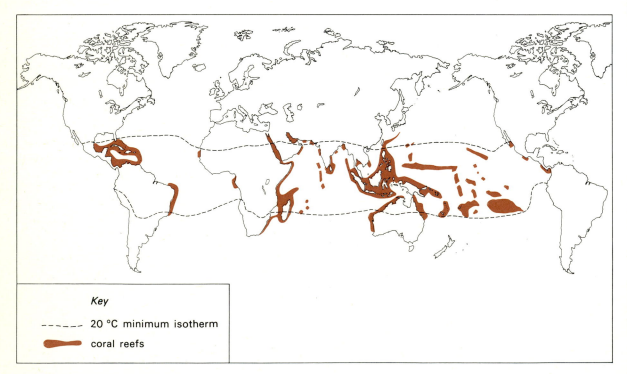

Figure 5.11 World distribution of coral reefs.

important in explaining why some deep water sediments are several thousand metres thick and yet still lie in deep water.

5.4 Coral reefs

We end this chapter with a brief examination of coral reefs and islands, partly because they are a special form of sedimentation and partly because to a considerable extent they depend for their present shape on sea level changes.

Coral reefs are structures of colonial corals living in a symbiotic relationship with a variety of algae. The skeletal material of the reef is calcium carbonate, extracted from solution in sea water by the organisms, which are therefore fixing the products of the terrestrial weathering system. Coral itself is quite capable of tolerating a fairly wide range of conditions, but the algae are sensitive to environmental conditions, and the distribution of coral reefs is therefore restricted by a number of factors. The more important of these are:

(a) *Temperature*. Reefs may form within waters where the minimum temperature is 16 °C, but it is generally agreed that prolific reef growth is limited to those areas where minimum temperatures are around 26 °C (Fig. 5.11).
(b) *Light*. The algae require adequate light for photosynthesis; and although 90 m is the theoretical depth limit for sufficient light, prolific growth is restricted to the 0–20 m zone.

(c) *Water quality*. Coral reefs develop poorly if salinity is too low or sediment content too high, and this tends to restrict growth near major river systems.

Within the limits imposed by these factors, coral reefs form important surface features in many tropical waters (Fig. 5.11). It was Charles Darwin who first recognised three main types of reef; and although more elaborate classifications are often used today, Darwin's scheme will still suffice as a starting point. In his voyage across the Pacific, he identified the following:

(a) *Fringing reefs* – platforms of coral growing out from land. The sediment and salinity problems mean that fringing reefs are less common along large landmasses, but they can be found on virtually all islands lying within the main occurrence zone (Fig. 5.12).
(b) *Barrier reefs* – reefs that lie at some distance from the coastline. They occur around many Pacific islands (e.g. Vanua Levu in the Fiji group, Fig. 5.12), but the most extensive, and best known, is the Great Barrier Reef, which lies off the Queensland coast of Australia.
(c) *Atolls* – subcircular reefs enclosing a lagoon with no other landmass visible. Many examples

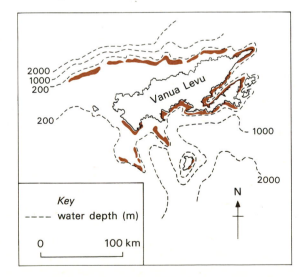

Figure 5.12 Vanua Levu, Fiji, showing fringing and barrier reefs.

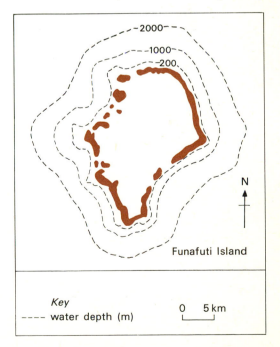

Figure 5.13 Funafuti Island, an atoll in the Ellice group.

(a) Fringing reef

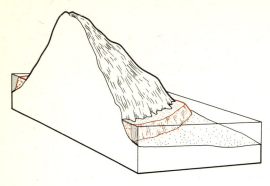

(b) Barrier reef

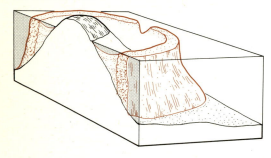

(c) Atoll

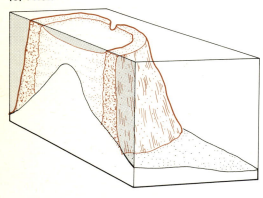

can be found in the Pacific (e.g. Funafuti Island in the Ellice group, Fig. 5.13).

Deep borings have shown that barrier reefs and atolls are founded, like fringing reefs, upon solid rock, which is usually volcanic in origin. As Darwin first realised, however, the base of such a reef is far deeper below the surface than should be possible given the light requirements of reef organisms. Darwin felt that the answer lay in the gradual subsidence of the underlying volcanic peak in such a way that coral continues to build up to the surface, so that a fringing reef will develop into first a barrier reef and later an atoll (Fig. 5.14). The stages of development suggested here are probably still quite acceptable, only the cause is now generally thought to be the postglacial eustatic rise in sea level (Fig. 5.6) observed earlier. Many of the Pacific coral islands identified in Figure 5.1 are thus a product of both internal and external processes. The original islands were basalt volcanoes formed by hot spots beneath the moving plate. Around each island grew a fringing reef, which developed into more complex forms as the rising sea level buried the original peaks.

Fig 5.14 Darwin's model of the evolution of barrier reefs (b) and atolls (c) from fringing reefs (a) as a result of relative sea-level rise.

Chapter 6

Destructive Plate Margins

In Chapter 4 we examined those tectonically active belts of the Earth (e.g. ocean ridges) where heat flow is higher than average, and we established that these features can be interpreted as zones of upwelling convection within the mantle. In this chapter we shall look at the other tectonically active belts of the Earth. In most cases these remaining areas include a deep ocean trench where heat flow is lower than average. Such a trench may be adjacent to either a volcanic island arc or a mountain chain. In a few instances a tectonically active mountain belt exists without any adjacent ocean trench, and these areas will also come within the scope of this chapter.

6.1 The ocean trench – island arc system

The ocean trenches of the world are almost all found along the edges of the Pacific Ocean (Fig. 1.1). Those on the northern and western margins of the Pacific lie adjacent to volcanic island arcs. Outside the Pacific the only true trench – island arc systems are those of Indonesia, the West Indies and the South Sandwich Islands.

Figure 6.1 shows part of the Tonga island arc and trench in rather more detail, but most of the characteristics found here apply equally to any other trench – island arc system. The trench itself is a long narrow feature, stretching for about 1200 km before grading into the Kermadec trench to the south. The maximum depth of the Tonga trench is 10 882 m. Eastwards, the floor of the trench rises to the relatively level surface of the Pacific seabed between 4000 and 6000 m. Westwards, there is a steep slope along the trench side as it rises to become the narrow ridge that breaks surface to form the Tongan islands. Each island is a volcanic cone with surrounding coral reefs. West of the islands the sea is less deep, giving a platform not more than 3000 m below sea level.

Apart from this distinctive topography, trench – island arc systems have several other features worthy of note. First is the surface geology. Islands like Tonga are founded upon volcanic rocks that are not of basalt composition. The distinctive lava type in this instance is **andesite** (Appendix A.2), which tends to have more sodic feldspar than basalt and includes more amphibole minerals and less pyroxenes than basalt (Appendix A.1). Second, there are the subsurface data. We have already noted that the heat flow pattern (Fig. 3.9d) indicates a distinct low over the trench as well as a high over the island arc. The seismic data are also of interest. As Figure 6.1c suggests, earthquake foci are found at increasing depths moving from the trench towards and beyond the island arc. The foci lie in such a tight linear grouping that this has been called the **Benioff zone** after its discoverer.

The interpretation of these data requires that we once again consider the possible movement of the lithosphere plates. In Figure 6.2, a hypothetical model for the Tonga system is suggested. In this model, the lithosphere of the Pacific plate is moving westwards and is forced to buckle beneath the piece of lithosphere that underlies the Fiji Platform. The trench is a natural result of this buckling. It is assumed that the descending slab of lithosphere, being chilled by contact with the surface, disrupts the subsurface isotherms (Fig. 6.2). The result is the initiation of a descending tongue of cooler material, which acts rather like the countercurrent to the rising convection of a spreading ridge. Low heat values at the trench reflect this subsurface cooling.

The locations of earthquake foci as predicted by Figure 6.2 vary somewhat. At the top of the descending plate earthquakes are generated along the upper surface by friction as a result of opposing plate motion. As the plate descends, however, it heats up, and the upper surface probably soon starts to melt, producing a sliding surface. Beyond this point most earthquakes are confined to the still solid centre of the descending plate, where stress is continuously being set up. At some point the whole of the descending plate reaches the same temperature as the surrounding mantle and is then thermally indistinct. This assimilation seems to be accompanied by the cessation of earthquakes. As shown in Figure 6.2, no earthquakes are recorded below 700 km, so that presumably marks the lower limit of the descending slab. It is notable that, although the Benioff zone extends to *less* than 700 km beneath some trenches, nowhere does it seem to *exceed* that figure. It is also significant that the

depth 700 km marks the lower limit of the asthenosphere.

The gradual heating of the descending slab has one other important effect. As temperatures rise at the upper surface of the descending plate, a partial melting of the oceanic crust occurs. Despite the basalt-type composition of this crust, partial melting allows the more volatile components to separate off, producing a magma that is of andesite rather than basalt composition. The melted material is less

(a)

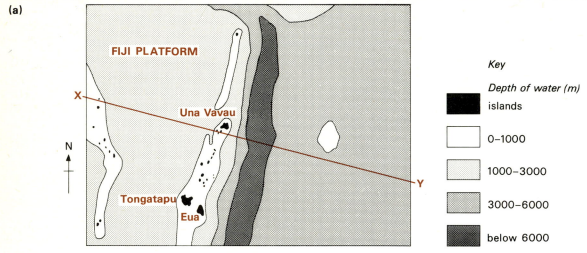

(b)

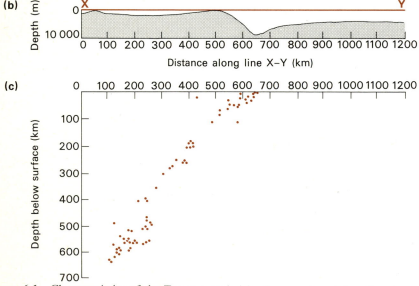

Figure 6.1 Characteristics of the Tonga trench–island arc system: (a) surface topography; (b) cross-section along the line X–Y; (c) observed earthquake foci beneath the island arc (after Isacks, Oliver & Sykes 1968).

dense than its surroundings and rises to produce the volcanic islands with their relatively high heat flow. The whole arrangement is illustrated in Figure 6.3.

The surface and subsurface evidence of the trench–island arc system may therefore be interpreted to show the reabsorption of oceanic lithosphere into the Earth's mantle. This process is often referred to as the **subduction** of the lithosphere, and the whole area is called a **subduction zone**.

Turning to the Pacific plate as a single unit, Figure 6.4 indicates the main tectonic features bordering the plate. Based upon the interpretation of the evidence presented so far, it seems clear that the Pacific plate is created along its southern and eastern margins by the spreading ridge that stretches from south of New Zealand round to the western coast of Canada. From this ridge the plate is moving to the northwest. Virtually all the islands on the surface of the plate are members of basalt ridges (e.g. the Hawaiian group) created by the passage of the plate surface across a stationary hot spot (see Section 5.1). The plate is subducted along its western and northern boundaries in a belt marked by a discontinuous series of ocean trenches, of which the Tonga–Kermadec trench is simply the most southern. A schematic view of an ocean plate, from spreading ridge to subduction zone, is given in Figure 6.5.

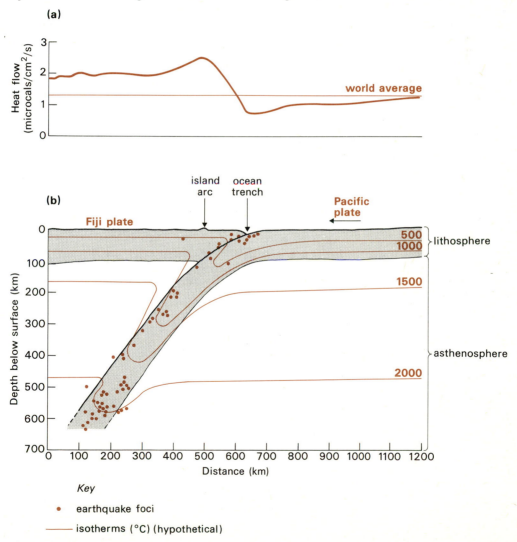

Figure 6.2   The Tonga trench–island arc system: (a) generalised heat-flow graph; (b) model of lithosphere subduction beneath the Tongan islands, indicating the possible distribution of temperature as shown by isotherms.

In practice, the subduction belt of the Pacific is complicated in a number of ways. The line of subduction is not itself continuous, for example. In particular, some kind of discontinuity seems to exist between the northern end of the Tonga trench and the southern end of the Marianas trench (Fig. 6.4). This may mark the site of a subduction-zone transform fault of the type illustrated in Figure 6.5. The rather confused area immediately south of this transform fault, which includes the small New Britain and New Hebrides trenches, probably marks localised subduction of northward-moving fragments of the Australian plate.

The rest of the western margin, from the Marianas to the Kurils and on to the Aleutians, consists of three main arc-shaped trenches backed by andesite island groups. The arc shape seems to be simply the form produced when a section of a sphere is bent downwards. (To make the point clearer, try depressing the surface of a table tennis ball with your finger; you will produce an arc-shaped fold, which is concave in the direction of downward movement.) Unlike the Tonga–Kermadec trench, where oceanic crust is subducting

beneath oceanic crust, the Pacific plate is subducting beneath continental crust along the Kuril–Japan trench (Fig. 6.6). South from here the situation is complicated by a double subduction belt. The Pacific plate subducts beneath the oceanic crust of the small Philippine plate, which in turn subducts beneath the continental crust of China along the Ryukyu trench – island arc system. A cross-section of the Ryukyu system (Fig. 6.7) shows that the island arc is simply appended to the edge of the continental crust, although it remains separated by a shallow marginal basin from the land area. Northwards, however, the Ryukyu arc grades into the Japanese Islands. The four main islands, Hokkaido, Honshu, Shikoku and Kyushu, are much larger than the islands of normal island arcs and structurally much more complex, consisting not only of contemporary volcanic rocks but also of older folded sediments and various igneous intrusions, which collectively form a mountain chain. A glance back at Figure 1.1 will perhaps suggest to the reader that mountain chains and island arcs are both parts of one linear system – a thought that takes us on to the next section.

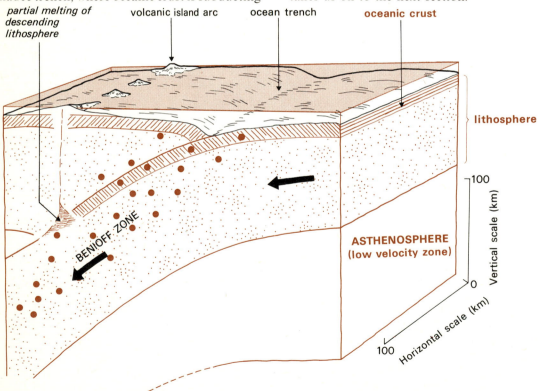

Figure 6.3 Block diagram to show the generalised relationships between subducting lithosphere, Benioff earthquake zone and partial melting of the lithosphere to produce andesite volcanoes at an island arc.

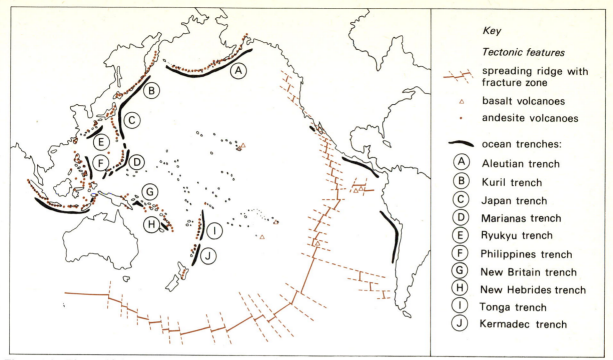

Figure 6.4 The Pacific plate with its active tectonic features. The basalt volcanoes of the spreading ridge and plate surface stand in contrast to the andesite volcanoes of the subduction zone.

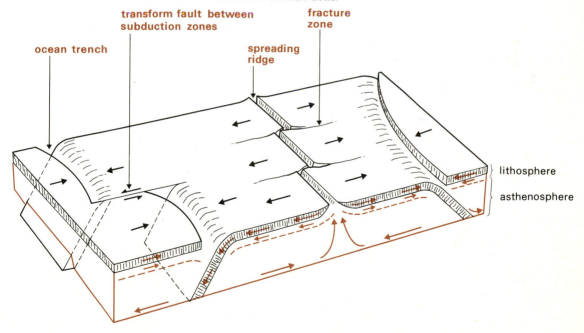

Figure 6.5 Model of a complete ocean-plate system from spreading ridge to subduction zone (after Isacks, Oliver & Sykes 1968).

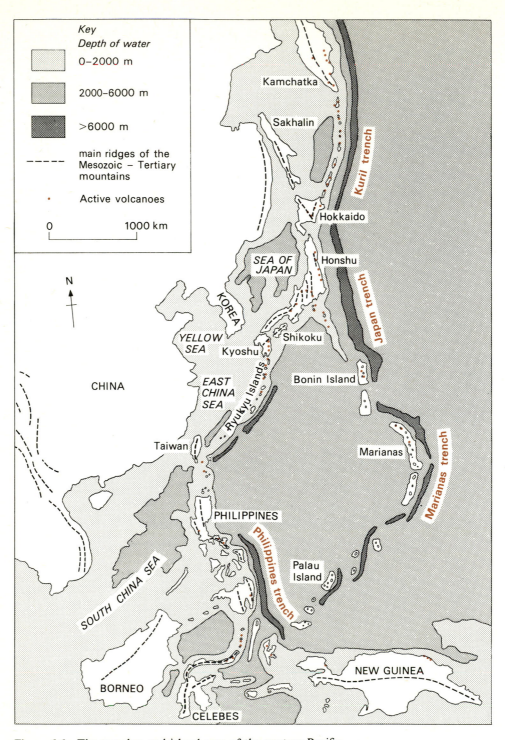

Figure 6.6   The trenches and island arcs of the western Pacific.

6.2 Mountain chains

The mountain chains shown in Figure 1.1, although including most of the world's highest peaks, do not include all those areas which may loosely be described as mountainous. The figure excludes, for example, the Appalachians of North America, the Scandinavian mountains, the Drakensbergs of South Africa and most of the mountain zone of Mongolia. Those areas shown as mountains in Figure 1.1 have in common their age of formation, since, although they include rocks of many different geological periods, all have been created within the last 200 million years and most within the last 30 million years or so. These so called 'young fold mountains' are therefore a product of the Mesozoic and Tertiary eras, whereas the other mountain areas of the world are thaeremnants of older epochs: the Palaeozoic and Precambrian.

The Mesozoic–Tertiary mountain systems fall roughly into three elongated belts:

(a) the *Rocky Mountains–Andes* chain, lying parallel to the western seaboard of the Americas;
(b) the *Alpine–Himalayan* chain, running eastwards from North Africa through the Mediterranean and Turkey to the Hindu Kush and Himalayas proper, before turning south through Burma–Thailand, the Malay peninsula and on to Indonesia;
(c) the *western Pacific* belt, which is a less continuous system of continental mountains and offshore islands running from the Kamchatka peninsula in the north through Japan and the Philippines to New Guinea.

In most cases these mountain chains consist of a number of parallel ridges, frequently grouped into two marginal mountain zones leaving an area of relatively low surface relief between to form a plain or plateau. The rocks that make up the mountain chains are usually somewhat older than the mountains themselves and are often of marine origin. They may include volcanic rocks (e.g. lavas, ash), which are frequently of andesite-type composition and were originally produced in submarine eruptions or possibly in island arc situations. Most of the rocks, however, are of sedimentary origin and include enormous thicknesses of deep water deposits (e.g. mudstones, greywackés, Appendix A.4).

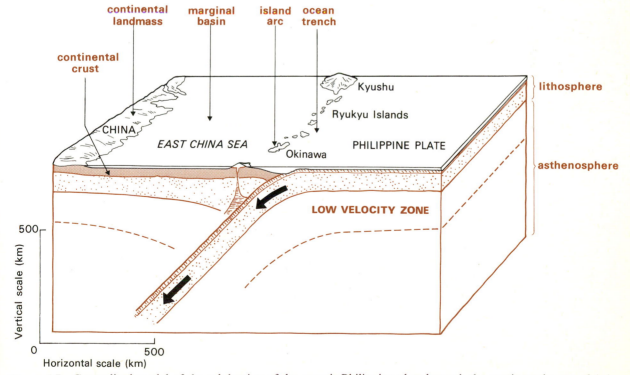

Figure 6.7 Generalised model of the subduction of the oceanic Philippine plate beneath the continental crust of Asia along the Ryukyu island arc (partly after Toksoz 1975).

The word **geosyncline** was introduced in the nineteenth century to describe a deep elongated trough in which such sediments may accumulate. The enormous thickness may be explained by assuming isostatic depression of the floor of the geosyncline under the weight of sediment. Today it is more common to assume that the deep water sediments accumulated in an environment such as the deep water margin of a continental area (Fig. 5.5), for which the term **eugeosyncline** (or eugeocline for short) is used. It is by no means certain that isostatic depression alone can account for some of the great thicknesses of sediment accumulated in a eugeocline, but this is a point to which we shall return. Most mountain chains, at least along their lateral margins, include some shallow water sediments (e.g. limestones), which probably formed on a continental shelf. Such deposits form elongated lenses, and their environment of deposition is termed a **miogeosyncline** (miogeocline), which, although of similar lateral dimensions to a eugeocline, is not likely to accumulate any great thickness of sediments.

The processes of mountain-building (**orogenesis**) involve the conversion of marine sediments and volcanics into upstanding elongate ridges. If we ignore the 'cause' of orogenesis for the moment, we can at least see the results of orogenic processes in the alteration of those original rocks. In virtually all mountain belts it is possible to observe the combined effects of those processes in terms of folding and faulting, igneous activity and metamorphism.

6.2.1   Folding and faulting

To some extent most rocks will deform under stress, with the result that the layers buckle or **fold**. Stress applied in a vertical direction will produce upfolds (**anticlines**) or downfolds (**synclines**, Fig. 6.8a). Most mountain belts consist of many parallel lines of folds, like corrugations, which can scarcely be due to simple vertical stress but are much more likely to be the result of lateral compression of strata over a wide area. In such fold belts the folds themselves are rarely simple open anticlines and synclines but consist of complex overfolding (**recumbent folds**) due to extreme lateral compression (Fig. 6.8a).

Under some stress conditions rocks are unable to deform sufficiently and therefore fracture or **fault**. We have already encountered several varieties of fault in earlier sections. In Chapter 4, for example, we examined the vertical faults bounding the sides

of oceanic and continental rift valleys and concluded that they have normally been produced by crustal tension, although, as Figure 4.12 indicates, the same result may be produced by compression. Mountain ranges often consist of extensive downfaulted blocks (**grabens**) or upstanding fault blocks (**horsts**) as well as minor fault escarpments produced by vertical movement (Fig. 6.8b). Similarly,

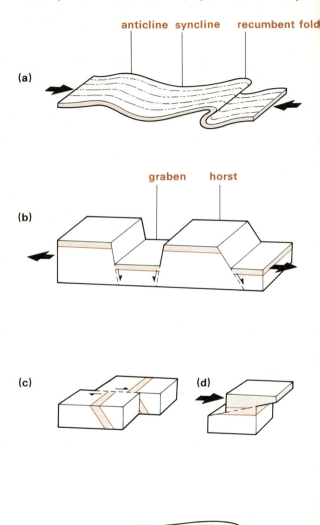

Figure 6.8   Folds and faults: (a) common types of fold; (b) vertical faults and the creation of fault blocks; (c) a transcurrent fault; (d) a thrust fault; (e) a nappe sheet, created by thrust faulting on a recumbent fold.

the transform faults of spreading ridges and subduction zones are special cases of what is often called a tear or **transcurrent fault**, which may be produced by opposing lateral stress at the surface (Fig. 6.8c). The most distinctive feature of mountain faulting, however, is the **thrust fault**, produced when one block of rock is pushed laterally across another (Fig. 6.8d).

Extreme lateral compression during orogenesis may lead to rock sheets travelling for many kilometres across the top of other strata either in very mobile recumbent folds or in thrust faults. These structures are called **nappes** (Fig. 6.8e).

6.2.2 Igneous activity

Apart from older submarine volcanic strata, most mountain chains contain igneous rocks that are contemporary with the folding and faulting of sediments. Two types of igneous activity are particularly common:

(a) Active volcanoes, mainly of andesite-type composition, usually accompany orogenesis and appear as a line of cones following the main fold ridges.

(b) Large masses of granodiorite or granite are intruded into the mountain core beneath the surface. These intrusions seem to originate at great depth as rising 'bubbles' of liquid material called **plutons**, which make their way towards the surface through the overlying layers of cool **country rock**. The plutons continue to rise by assimilating the country rock or by pushing it to one side, but eventually they come to a halt while still at some depth, where they solidify. Intrusions may adopt a number of distinctive forms (Appendix A.2), but by far the most common is the shapeless mass called a **batholith**. Extensive batholiths, elongated along the direction of fold axes, can be found underlying most mountain cores where they have been exposed by later erosion. Given the rapid denudation that accompanies orogenesis, major intrusions like batholiths may continue to rise isostatically long after the original intrusion, until they actually form the peaks of the mountain range. Everest, it may be noted, is an eroded peak of such an igneous intrusion.

6.2.3 Metamorphism

If rock material is subjected to heat and/or pressure after the time of its original formation, it may undergo a variety of textural or mineralogical changes collectively known as **metamorphism**. The processes of orogenesis inevitably produce a range of conditions in which metamorphism can occur. Although it is sometimes rather difficult to separate them in practice, three types of metamorphism are usually recognised:

(a) **Dynamic metamorphism.** The application of pressure, or stress, to rock material may cause fragmentation of the rock. Alteration of this type may accompany faulting, and it is particularly common to find the surface of a thrust fault coated with recrystallised rock powder called **mylonite**.

(b) **Thermal** or **contact metamorphism.** The heat given off by a cooling igneous mass can cause alteration of the country rock with which it is in contact. Lava flows will produce only a small scale metamorphism of the underlying surface, but a large batholith may cause alteration across a zone several kilometres wide in all directions – a **metamorphic aureole**. Whereas thermal metamorphism is unlikely to produce great changes in older igneous rocks, themselves produced by high temperature processes, sedimentary rocks may show considerable alteration. Fine-grained rocks (e.g. mudstones in particular) may be baked to a hard flint-like material called **hornfels**, which is far more resistant to erosion than the original rock.

(c) **Regional metamorphism.** In almost all mountain areas of the world considerable alteration of both sedimentary and igneous rock seems to occur over a wide area as the result of the simultaneous application of heat and pressure. Such regional metamorphism usually shows a pattern of most intense alteration towards the centre of the orogenic belt, with surrounding zones of decreasing metamorphic intensity. Some of the extremely high temperature and high pressure conditions required for regional metamorphism can be explained by the lateral compression that occurs during folding, but it seems fairly clear that the most intense forms of metamorphism require that the mountain cores be originally buried at much greater depth within the Earth's crust. An enormous range of rock types can be produced by regional metamorphism, since the nature of any one particular piece of rock will depend not only upon its original composition but also upon the exact temperature and pressure conditions reached during metamorphism. The margins of mountain belts often include the altered products of shallow water sediments (e.g. **quartzite** from sandstone, **marble** from limestone, **slate** from mudstone). Towards the centre of a mountain chain, however, the extent of alteration increases, and sedimentary and igneous rocks alike become part of a sequence of **schists** and **gneisses**. Schist and gneiss are most commonly of granite-type mineral composition.

6.3   Mountain chains and the subduction of the lithosphere

Having noted some of the common characteristics of the world's orogenic belts, we shall turn our attention to explaining how the processes of orogenesis are produced. We shall start by examining two specific areas of recent mountain-building activity.

6.3.1   The Andes

The Andes (Fig. 6.9) follow the entire western seaboard of South America, where they consist of a series of ridges lying parallel to the coast and reaching heights of over 7000 m. In the central section, the ridges separate into the clearly defined Western and Eastern Cordilleras, between which lies the high Altiplano of Bolivia, where some of the world's highest lakes are found.

At the present day this is a tectonically active area. From the clearly defined Peru–Chile trench, which lies immediately offshore, a Benioff zone of seismic activity descends beneath the mountains. Coupled with the location of active volcanoes of andesite-type composition (the term 'andesite' comes from the Andes) along the main ridges, the evidence obviously indicates that this is a subduction zone. Reference back to the tectonic map of the Pacific (Fig. 6.4) suggests that eastward spreading from the Easter Island Ridge has produced the small Nazca plate, now forced to subduct beneath the South American plate. Figure 6.10 gives a much simplified view of this subduction zone in cross-section. The active volcanoes of the Andes occupy the same relation to trench and subducting plate as an oceanic island arc, except that they are somewhat mixed up with an orogenic belt.

The creation of that orogenic belt, however, was probably a fairly complex affair. In the central section of the Andes (Fig. 6.10) the subsurface geology reveals that the western ridges are largely composed of volcanic rocks of Mesozoic age

(240–65 million years old). These rocks seem to have formed in an offshore marine environment such as an island arc, although they have subsequently been subjected to intense folding, faulting and metamorphism. By contrast, the eastern ridges contain deep-water sedimentary rocks, which are much older, being of Palaeozoic age. These rocks have also been much altered by intense folding, faulting and metamorphism, although they grade eastwards into other Palaeozoic sediments, which are of continental shelf origin and have suffered far

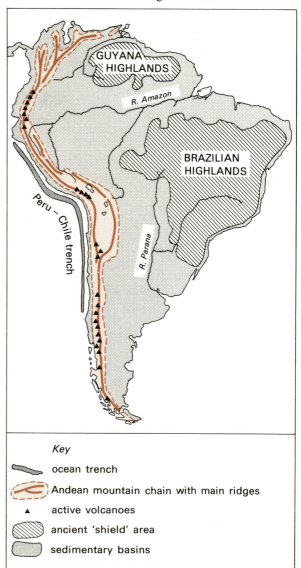

Figure 6.9   South America, showing the structure of the Andes and its relationship to the Peru–Chile trench and the exposed 'shield' areas of the continent.

less intense alteration. Granodiorite batholiths have been intruded into the mountains at numerous places, but these are mainly Tertiary in age and seem to become progressively younger to the east. The high plateau between the ridges is underlain by a thin veneer of recent sediments.

This evidence clearly suggests that the Palaeozoic sediments accumulated over a long period of time as eugeoclinal and miogeoclinal lenses along the stable margin of a continent. The volcanic rocks seem to have originated in an offshore island arc. If eugeocline and island arc came into collision, mountain-building would commence, but at the moment island arc and eugeocline seem to be on the same plate and therefore cannot move into collision. (This is the present situation illustrated for the Ryukyu arc in Fig. 6.7.)

One *possible* model that overcomes this difficulty is illustrated in Figure 6.11. Originally, subduction may have operated in the opposite direction, with oceanic portions of the South American plate subducting westwards beneath the Nazca plate (Fig. 6.11a). Sediments accumulated along the

western margin of South America for some time before active subduction initiated an island arc on the Nazca plate. Eventual collision between the continental margin and the island arc (Fig. 6.11b) would create the extreme lateral compression required to fold and fault both volcanic and sedimentary rocks into mountain ridges. Miogeoclinal deposits, being underlain by rigid continental crust, would suffer less extreme compression than eugeoclinal sediments. Given that continental crust is less dense than oceanic crust, continuing subduction would probably cause the oceanic 'leader' of the South American plate to become detached from the more buoyant continent (Fig. 6.11c). Nonetheless, material caught within the orogenic belt might be dragged downwards into a higher temperature environment, thus assisting the processes of metamorphism. Assuming that spreading continued from both South Atlantic and Easter Island Ridges during this time, a new subduction zone would have to be created, and it seems reasonable to suppose that the denser oceanic crust of the Nazca plate would be forced beneath the

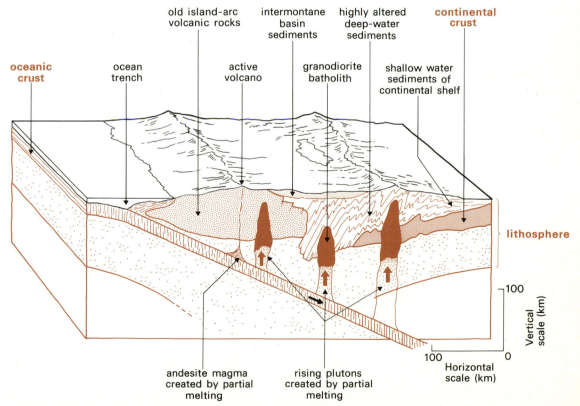

Figure 6.10 Generalised block diagram illustrating the present-day tectonic relationships in the central section of the Andes.

mountains to create the present tectonic pattern (Fig. 6.11d).

Partial melting of oceanic crust from this descending plate would provide the material for the modern volcanoes along the mountain chain. It

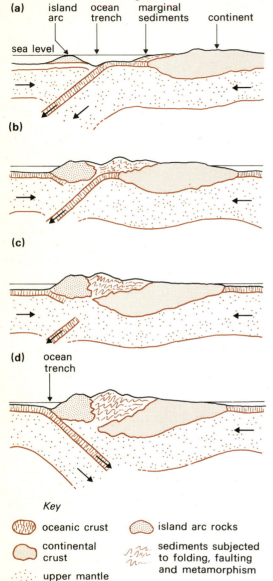

Figure 6.11 Model of continent–island arc collision (after Dewey & Bird 1970): (a) island arc and continent are on separate but advancing plates; (b) collision creates a mountain chain from folded sediments and island arc volcanics; (c) leading oceanic section of the subducting plate becomes detached from the more buoyant continental crust; (d) continued spreading from ocean ridges causes a new subduction zone to open up beyond the margins of the mountains.

seems likely that the batholiths intruded into the mountains are also part of the modern tectonic pattern, with plutons created initially by the same partial-melting process (Fig. 6.10). It is quite possible, however, that the material of which the plutons are made is partially derived from rocks of the orogenic belt dragged downwards to a depth where melting begins to occur. The decreasing age of batholiths eastwards presumably reflects the eastward progress of the present subducting plate.

The recent sediments occupying the surface of the high plateau seem to have been produced since orogenesis started, by the denudation of the surrounding mountain peaks. The coarse sediments laid in a mountain basin or along the foothills of a mountain chain as a result of post-orogenic denudation are common in all mountain areas and are normally known by their French name of **molasse**.

6.3.2   The Alps

The Alps are part of a more complex mountain system running through southern Europe. As Figure 6.12 suggests, although the whole orogenic belt is elongate, it is also twisted to a considerable extent, giving embayments (e.g. the Adriatic basin) outside the system proper. Within the rather confused pattern of structural ridges, it is possible to discern a northern ridge group (the Sierra Nevada, Pyrenées, Alps, Carpathians, Caucasus Mountains and on to the Hindu Kush) and a southern ridge group (the Atlas Mountains, Apennines, Dinaric Alps, Taurus Mountains, Zagros Mountains and on to the Himalayas). In places the gap between the two ridge groups is wide and is occupied by a variety of surface features (e.g. the Anatolian Plateau, the Hungarian Plain, the Tyrrhenian Sea).

The ridge systems become one in the Alps themselves, where lateral compression has been so extreme that the rocks have been squeezed like toothpaste from a tube into a series of gigantic nappes, which have been pushed to the northwest (Fig. 6.13). Only in the southeast of the Pennine Alps are the actual roots of the folds seen. To the northwest, peaks such as the Matterhorn (4478 m) are the eroded remnants of these nappes formed in schist and gneiss, which seem to represent altered eugeoclinal sediments, and related granitic intrusions from an early stage of orogenesis. The high calcareous nappes of the Bernese Oberland probably represent something closer to miogeoclinal deposits and include limestones now altered to marble. Within this zone, shattered relics of underlying continental margin crust have been up faulted and form upstanding granitic blocks (e.g. the Mont Blanc massif). On the northwestern edge of the Alps some of the nappes have actually travelled across the post-orogenic molasse of the Swiss Plain. Beyond this Tertiary basin a further, although much less altered, mountain ridge exists where Mesozoic miogeoclinal sediments have been folded into the Jura Mountains.

As far as the present tectonic situation is concerned, the Alps do not lie parallel to any recognisable ocean trench as do the Andes. The Mediterranean basin is shallow and is crossed by the orogenic belt. Similarly, although this is a seismically active area, at least along the southern margin (in the Apennines, Dinaric Alps, Greek Islands and Taurus Mountains), most of the earthquakes have a shallow focus, and there is little to indicate the existence of a Benioff zone. Finally, volcanic activity is also somewhat restricted, being confined to the Appenines and Aegean Sea (Fig. 6.12), although this does include some of the world's most famous volcanoes. The area does therefore contain some elements of a subduction zone but little to suggest active subduction at the present day.

On the other hand, the completely independent evidence for continental drift examined in Chapter 2 indicates that Africa and Europe have moved relative to one another in the past. In Figure 2.6, for example, an arm of the ocean called the Tethys Sea, which was in existence 200 million years ago during the early Mesozoic, was subsequently closed by the anticlockwise rotation of Africa. Sediments laid down in the Tethys would have been caught between the two continental masses of Africa and Europe. Closure of the Tethys Sea implies that the subduction of lithosphere bearing oceanic crust must have occurred. Figure 6.14 is an attempt to show how two continents may come into collision as the result of subduction along one of their margins. In this model orogenesis, involving the folding, faulting and metamorphism of eugeoclinal sediments, would occur without the need for an island arc, although volcanoes and igneous intrusion would probably develop above the subducting plate. Once the continents had collided, closing the ocean and eliminating the ocean trench, subduction would probably cease, as the subducting ocean plate would again become detached from the less dense continents and orogenic mass, which resist

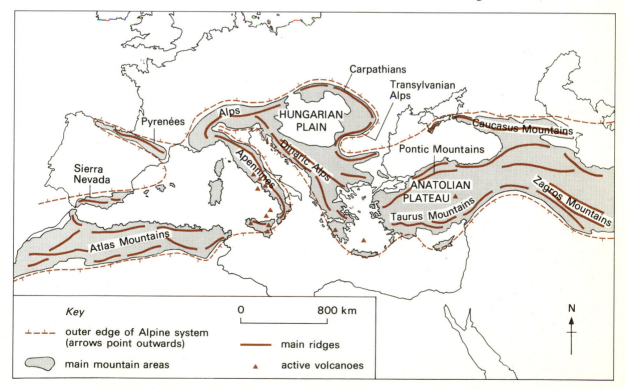

Figure 6.12 The Alpine orogenic belt.

being dragged downwards. It is possible that the Alps, with a low level of volcanic activity and mainly shallow-focus earthquakes, represent a fairly late stage in this sequence of events. Subduction may have largely ceased, and so it is perhaps preferable to regard this type of orogenic belt as a **collision zone** rather than as a subduction zone.

The Andes and Alps therefore represent two possible types of orogenic situation: the former the collision between a continent and an island arc, the latter the collision between two continents. In each case it is the subduction of lithosphere that brings together sections of the Earth's crust, which can act as a vice to crush the sediments deposited along a continental margin. In each case subduction leads to volcanic and intrusive activity contemporaneous with the mountain-building and may also encourage the high level metamorphism of the crushed sediments.

Regarding the rest of the world's mountain chains, the situation is not always as clear cut as this description may suggest. The Himalayas are probably a larger scale version of the double-continent collision seen in the Alps. In this case, however, the mountains were created by the northward movement of India against Asia.

The mountains of western North America initially appear to be a marginal mountain chain like the Andes, but closer examination reveals a rather complex history. Like the Andes, this mountain chain falls rather obviously into two main ranges: the Rocky Mountains in the east and a western range consisting of the Coast Mountains of Canada and the Cascades–Sierra Nevada of the United States. The Rocky Mountains, however, are considerably older, having been formed in the Mesozoic, and now show very little orogenic activity. The western ranges are of Tertiary age, but even here it seems that subduction in the fairly recent past has largely ceased. Volcanic activity is limited to Alaska and the Cascades, and earthquake activity, although very common, is largely confined to shallow focus events, which seem related to the transform faults of an active spreading ridge that can be seen in the Gulf of California (Ch. 4) and again off the Oregon coast (Fig. 6.4).

The discontinuous mountain fragments of the western Pacific are also something of a puzzle. As Figure 6.7 indicates, the subduction of the Pacific, or Philippine, plate beneath the continental crust of Asia results in an island arc resting on the edge of the continent and separated from the mainland by a marginal sea, making it impossible for continent and island arc to collide. Nonetheless, this same tectonic situation includes mountain zones like the Japanese Islands, which may therefore indicate orogenesis without actual collision. We shall examine this possibility in more detail in the next section.

6.4 The growth of continents and orogenic cycles

The creation of the Andes has in effect added to the extent of the South American continent. The South American continent east of the Andes (Fig. 6.9) consists largely of granite or granodiorite-type crystalline rocks of Precambrian age. These ancient

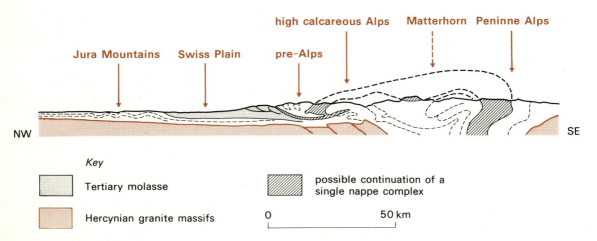

Figure 6.13 Highly generalised section across the Pennine Alps and Swiss Plain to the Jura Mountains. Only the simplest of the nappe fold systems is shown (partly after Holmes 1944).

rocks are exposed in the upland areas of the Brazilian and Guyana Highlands (the so called 'shield' areas) but underlie the later (mainly Palaeozoic) sediments that fill the Orinoco, Amazon and Paraguay basins. During the Andean orogenesis the Palaeozoic and Mesozoic sediments and volcanic that accumulated along the western edge of the continent were turned into crystalline material (metamorphic or igneous) and firmly welded on to the existing continental crust. In time the Andes will be reduced by denudation and possibly covered by a later veneer of sediment.

A rather more complex situation can be seen in North America. In Figure 6.15 areas are shown according to the age of orogenesis of the rocks

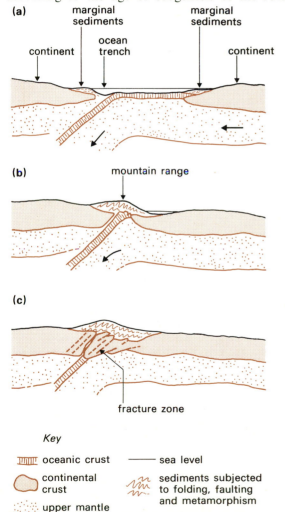

Figure 6.14 Model of double-continent collision, producing an orogenic belt (c) from two sets of marginal sediments (a) (after Dewey & Bird 1970).

exposed at the surface. In the north the Canadian shield is composed of rocks of Precambrian age, whereas along the eastern seaboard Palaeozoic rocks make up the Appalachian Mountains. Looked at over all, however, the pattern is one of the gradual accumulation of orogenic belts around the oldest rocks, which lie south and east of Hudson Bay, although the belts are not always in the right sequence (the Caledonian and Hercynian mountain belts cross in New York State). The word **craton** (Section 2.3) is often used to describe the oldest core area of a continent (e.g. the area around Hudson Bay). Examination of all the major continents shows that they consist of an ancient cratonic core surrounded by the remnants of orogenic belts of varying age, even where much of this basement material is overlain by later sediments.

Continental crust is therefore produced during orogenesis, and continents grow by the accretion of orogenic belts along their margins. As Figure 6.15 shows, however, one of the odd things about continental growth is that the same continental margin must alternately provide conditions of sedimentation and then of subduction. So far we have assumed that mountain-building occurs when sediments, accumulated on a stable continental margin, come into collision with an island arc or another continent as the result of lithosphere subduction. Repeated cycles of sedimentation and orogenesis on the same continental margin raise the possibility that sedimentation may *cause* orogenesis. Figure 6.16 indicates a possible model for this relationship, in which continued sedimentation and isostatic depression cause downwarping of the oceanic crust until it starts to subduct beneath the continent. The initiation of a subduction zone immediately beneath a eugeocline may provide sufficient lateral compression to cause orogenesis without the need for continent or island arc collision.

Corroborative evidence for the creation of a subduction zone by sedimentation comes from the geological match on the two sides of the North Atlantic. The Caledonian orogeny, which created the older part of the Appalachians (Fig. 6.15), also produced mountain zones in North Africa, Great Britain, Greenland and Norway (Fig. 2.2). There is ample evidence from the rocks of this period that eugeoclinal and miogeoclinal sediments were laid down on the margins of two continents (America–Greenland and Baltic–Africa) before subduction occurred, bringing about orogenesis. After the Caledonian mountains had been made the continents split apart again, although along a

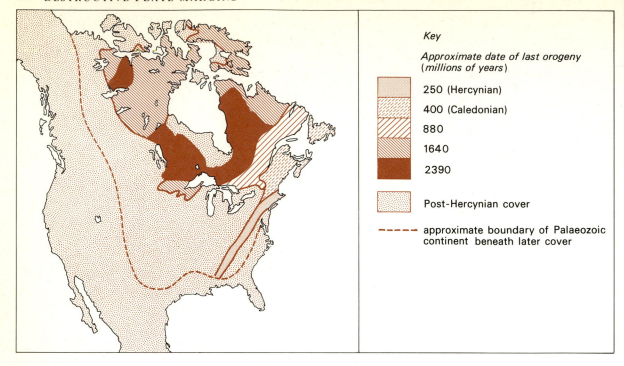

Key

Approximate date of last orogeny
(*millions of years*)

250 (Hercynian)

400 (Caledonian)

880

1640

2390

Post-Hercynian cover

– – – – approximate boundary of Palaeozoic
continent beneath later cover

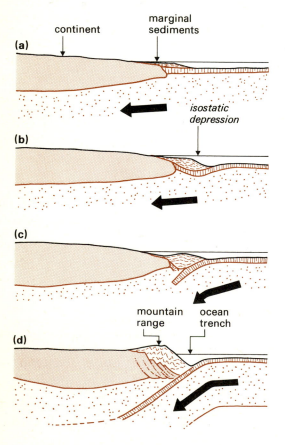

Figure 6.15   Orogenic belts of North America up to the end of Palaeozoic.

Figure 6.16   Development of subduction by sedimentation: (a) marginal sediments accumulate along the trailing edge of a continent; (b) sediments gradually cause isostatic depression of the underlying oceanic crust; (c) oceanic crust is forced down, until it starts to sink into the mantle, creating a subduction zone; (d) active subduction squeezes the sediments into a mountain chain.

slightly different line, leaving a northern continent comprising the Canadian shield, Greenland, northwestern Britain and the Baltic, and a southern continent based upon Africa. A new sequence of sediments was laid along the margins of this opening ocean, until once again subduction occurred, the ocean closed and the Hercynian mountains were formed, including the newer Appalachians, southwestern England, Armorica, the Ardenne–Rhine Uplands, the Massif Central and large sections of the Iberian Peninsula. For a third time the continents split apart, this time along the line of the present Atlantic, and at the moment sedimentation is occurring along the receding continental margins (see Ch. 5). Presumably, it is quite possible that at some future date the Atlantic will close yet again, creating another orogenic belt.

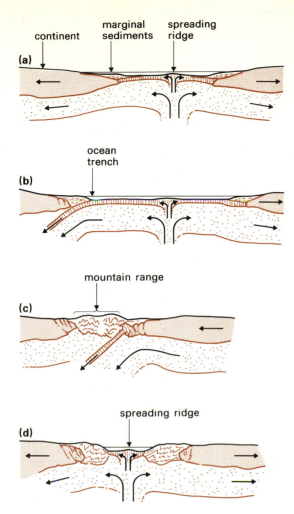

Figure 6.17 A possible orogenic-cycle model: (a) continents separate by spreading of the ocean ridge. Sediments accumulate along trailing edges; (b) isostatic depression by sediments on at least one margin creates a subduction zone; (c) active subduction on one continental margin closes the ocean, creating a mountain range; (d) another cycle starts with renewed spreading of the ocean ridge, which splits the continents apart again.

This repeated opening and closing of an ocean approximately along the lines of the present Atlantic obviously suggests the operation of some kind of cyclic mechanism. In Figure 6.17 a possible model for the cyclic behaviour of the Atlantic is based upon the idea that isostatic depression by sedimentation eventually causes subduction.

6.5 Orogenic belts of the British Isles

The accretion of orogenic belts on to the North American continent by cyclic opening and closing of the 'Atlantic' has been repeated on the European side of the ocean. As it happens, the British Isles, although covering only a very small area, include structures produced by at least three clear orogenic cycles (Fig. 6.18). Of these three cycles, the first two (Caledonian and Hercynian) involved orogenic cycles based upon the 'Atlantic', whereas the third (Alpine) is associated with the closing of the Tethys Sea between Europe and Africa, which produced the Alps. Since the site of the British Isles moved progressively away from the centre of orogenic activity during these successive cycles, the rocks of this country show an interesting range of features associated with mountain-building, from very intense folding, faulting and metamorphism to gentle tilting of marginal sediments.

By adding in areas where the rocks predate the Caledonian cycle and areas where sediments and volcanic activity are part of the present Atlantic cycle, we can describe the building of the British Isles in five stages:

(1) the pre-Caledonian stage;
(2) the Caledonian cycle;
(3) the Hercynian cycle;
(4) the Alpine cycle;
(5) the post-Alpine stage.

6.5.1 Pre-Caledonian areas

The rocks of the Outer Hebrides and northwestern Scotland (Fig. 6.18) consist of highly contorted schist and gneiss cut by innumerable dykes. These Lewisian rocks (named after the Isle of Lewis) probably represent very ancient sediments, which were altered during at least two major orogenic episodes during the Precambrian period. They are therefore similar to parts of the Canadian shield, to which they were presumably once attached. They therefore represent the edge of the continent that took part in the Caledonian cycle.

6.5.2 Caledonian cycle

During the late Precambrian and lower Palaeozoic the site of the British Isles was occupied by an ocean, with the ancient Lewisian continent lying to the northwest and another continent probably some way to the southeast. Fluvial conglomerates and sandstones were laid down on the edge of the Lewisian continent (the Torridonian rocks of

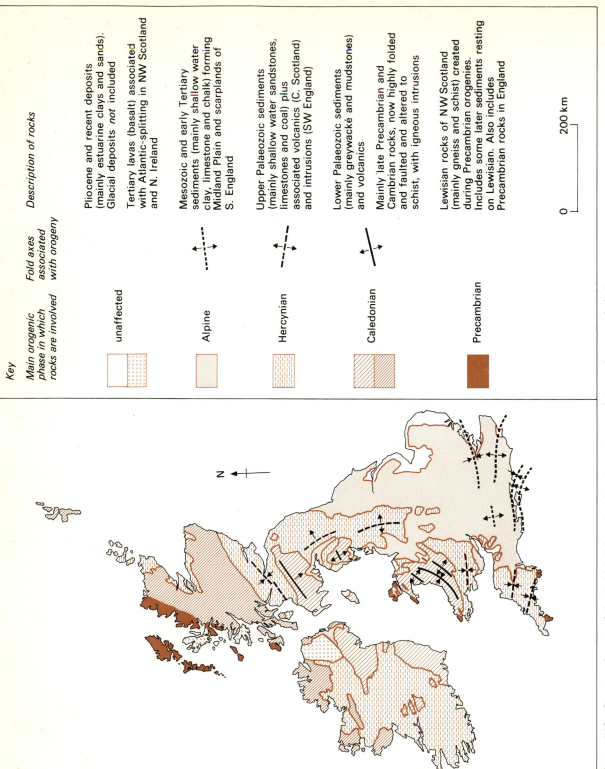

Figure 6.18 Structural map of the British Isles, showing the outcrop of the main orogenic belts.

Key

Main orogenic phase in which rocks are involved | *Fold axes associated with orogeny* | *Description of rocks*

unaffected — Pliocene and recent deposits (mainly estuarine clays and sands). Glacial deposits *not* included

Tertiary lavas (basalt) associated with Atlantic-splitting in NW Scotland and N. Ireland

Alpine — Mesozoic and early Tertiary sediments (mainly shallow water clay, limestone and chalk) forming Midland Plain and scarplands of S. England

Hercynian — Upper Palaeozoic sediments (mainly shallow water sandstones, limestones and coal) plus associated volcanics (C. Scotland) and intrusions (SW England)

Caledonian — Lower Palaeozoic sediments (mainly greywacke and mudstones) and volcanics

Mainly late Precambrian and Cambrian rocks, now highly folded and faulted and altered to schist, with igneous intrusions

Precambrian — Lewisian rocks of NW Scotland (mainly gneiss and schist) created during Precambrian orogenies. Includes some later sediments resting on Lewisian. Also includes Precambrian rocks in England

0   200 km

northwestern Scotland), and miogeoclinal shales and limestones were deposited in shallow water on the southeastern side of the ocean in what is now the Shropshire area. In the centre, along a trough aligned NE–SW, eugeoclinal mudstones and greywackés were formed. At some fairly early point during the cycle, subduction must have started within the eugeocline, since andesitic lavas and ash, typical of an island arc formation, are recorded in the Ordovician rocks of Snowdonia and the Lake District, where they often form the highest points in the landscape (e.g. Snowdon, the Glyders, Scafell). As the cycle progressed, subduction continued, and the ocean gradually closed.

Miogeoclinal deposits on the continental margins, such as those in Shropshire, were subjected to moderate tilting, and subsequent denudation has left the limestone–shale sequence as a scarp and vale landscape (e.g. Wenlock Edge). The eugeoclinal and island arc deposits were subjected to far greater compression as the ocean closed, however, and therefore display a wide range of alterations. Over much of central and northern Wales, the Lake District and the Southern Uplands of Scotland, the sediments were subject to intense folding. In Wales, for example, a complex series of anticlinal and synclinal structures (Fig. 6.19) was produced, their axes lying at right angles to the direction of compression. Throughout this belt fine-grained mudstones suffered mild metamorphism and are now seen as slates. In the Northern Highlands of Scotland, however, the effects were far more dramatic. In many places intense folding produced recumbent or even nappe folds, which have made the geology of the area difficult to unravel. The sediments must also have been dragged downwards into the subduction zone, as high grade metamorphism is widespread. Most of the rocks are schist and gneiss, although marble and quartzite also occur.

The extent of metamorphism in an orogenic belt like this is usually determined by the presence of **indicator minerals**, which form in different temperature–pressure environments. Mapping the metamorphic grade of the Scottish Highlands produces

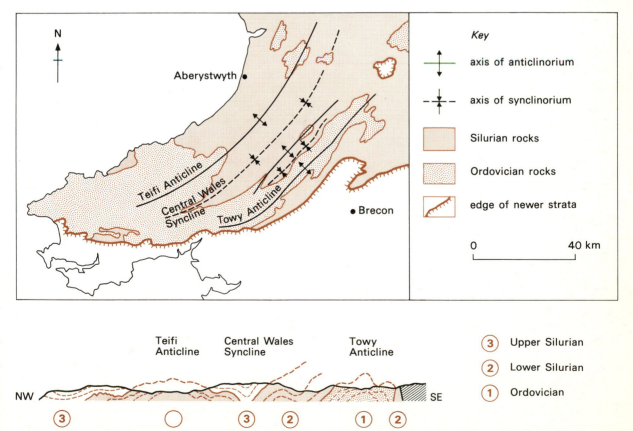

Figure 6.19 Caledonian fold structures in central Wales.

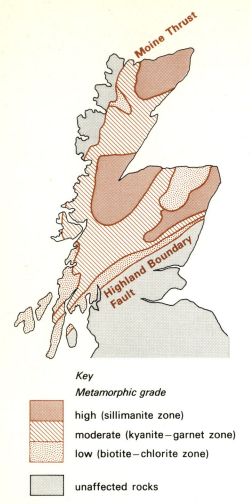

Key

Metamorphic grade

high (sillimanite zone)

moderate (kyanite – garnet zone)

low (biotite – chlorite zone)

unaffected rocks

Figure 6.20 Caledonian metamorphism of the Scottish Highlands, with metamorphic grade shown by indicator mineral assemblages. In this map the position of the area north of the Great Glen Fault is reconstructed on the basis of the assumed movement shown in Figure 6.21.

the sort of pattern shown in Figure 6.20. In this particular reconstruction the area north of the Great Glen is moved to the northeast, back to the position that it probably occupied at the time of orogenesis. On this basis it can be seen that most intense metamorphism was centred on the Moray Firth and Caithness areas.

At the time of maximum folding, granodiorite batholiths were intruded along the fold axes, and these are now exposed in some of the highest zones (e.g. the Grampians). Faulting seems to have been common throughout the Highlands. Apart from many minor faults, the Caledonian orogeny seems

to have initiated the Great Glen Fault, a tear fault (Fig. 6.21), which can now be traced through Loch Ness and Loch Lochy as well as the fault-bounded Scottish Central Valley, which lies between the Highlands and Southern Uplands. Perhaps most spectacular, however, was the development of a massive thrust fault (the Moine Thrust), which pushed metamorphosed eugeoclinal rocks of the Highlands (the Moine Series) northwestwards over Palaeozoic miogeoclinal sediments (Fig. 6.21).

6.5.3 Hercynian cycle

By the end of the Caledonian orogeny, mountains covered most of the area now occupied by the British Isles. Early in the Hercynian, a new ocean basin opened up to the south of the mountains, and marine sediments were laid in a belt that passes through Devon and Cornwall but stretches westwards into the southern margin of the North American continent and eastwards through Europe. Over the rest of the British Isles, the Devonian period was dominated by denudation of the mountains, although coarse sediments (the so-called Old Red Sandstone) accumulated in low-lying areas such as the coastal edge of the ocean, which passed through southern Wales, and in mountain basins (e.g. the Central Valley of Scotland and the area around the Moray Firth). The rift faults of the Central Valley continued to develop during this time, and they were associated with andesite volcanoes, which poured out lavas to form the Campsie Fells and Ochill Hills.

By the Carboniferous period, although deep water sedimentation of eugeoclinal type continued in Devon and Cornwall, producing mudstones and greywackés of the Culm Measures, the rest of the country had been reduced to a low surface by denudation. A rise in sea level therefore turned much of the British Isles into a continental shelf environment, within which the Carboniferous limestone was formed. Later changes in denudation on the adjacent landmasses filled the shallow water with great deltas of coarse sand (the Millstone Grit), and it was on the top of these deltas that the swamp conditions existed that eventually produced the Coal Measures.

The Hercynian cycle ended as the ocean to the south closed. The site of a subduction zone does not appear in the British Isles, but the eugeoclinal sediments of Devon and Cornwall experienced intense folding along an east–west axis and mild metamorphism of the mudstones to slate. More important, perhaps, was the intrusion along the line

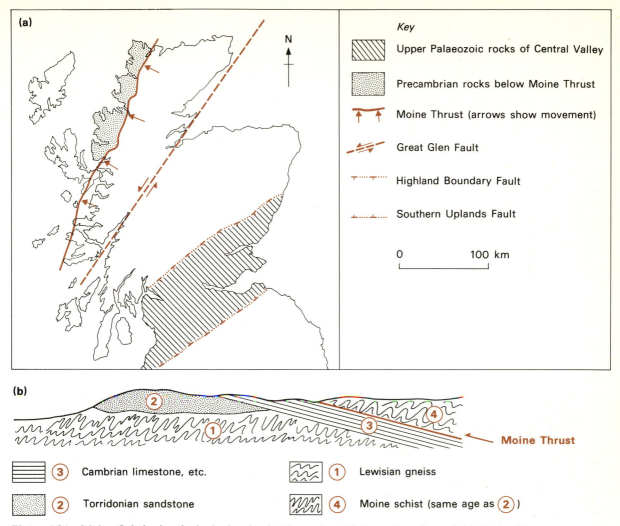

(a)

Key

Upper Palaeozoic rocks of Central Valley

Precambrian rocks below Moine Thrust

Moine Thrust (arrows show movement)

Great Glen Fault

Highland Boundary Fault

Southern Uplands Fault

0   100 km

(b)

Moine Thrust

3 Cambrian limestone, etc.

2 Torridonian sandstone

1 Lewisian gneiss

4 Moine schist (same age as 2)

Figure 6.21 Major Caledonian faults in Scotland. The section (b) shows the effect of the Moine Thrust.

of the fold axes of a major granite batholith (Fig. 6.22). The granite moors of southwestern England (e.g. Dartmoor, Bodmin Moor) are merely the upper exposed surface of a continuous batholith. Similar intrusions mark the line of Hercynian folding throughout the western Appalachians and Central Europe.

The miogeoclinal sediments covering much of the rest of Britain (Fig. 6.18) were still underlain by continental crust (of Caledonian age) and so were far less affected by orogenic activity. They were, nonetheless, subject to compression, which produced broad anticlinal and synclinal folds. In southern Wales a major syncline was formed around the Coal Measures, with the Old Red

Sandstone rocks on the northern edge creating the uplands of the Brecon Beacons. In northern England the rocks formed the Pennines along a north–south axis. The centre of this anticline is eroded in many places to reveal the Carboniferous limestone, which has been weathered into the famous **karst** areas of the Craven District of Yorkshire and the Peak District of Derbyshire. The high moors of the Pennines are formed by the overlying Millstone Grit, and the Coal Measures have been removed entirely from the top of the Pennines, so that they can now be found only along the flanks of the hills. Although these deposits suffered no generalised metamorphism, a certain amount of igneous activity occurred. In particular,

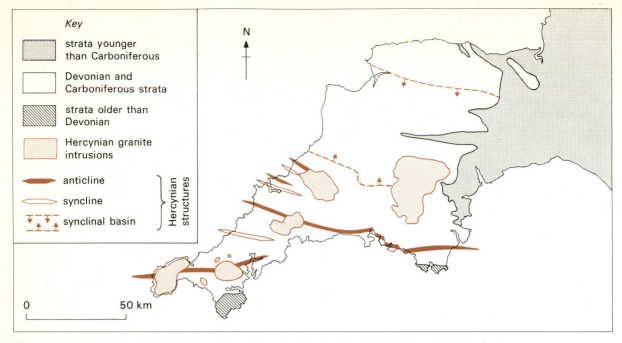

Key

- strata younger than Carboniferous
- Devonian and Carboniferous strata
- strata older than Devonian
- Hercynian granite intrusions
- anticline
- syncline
- synclinal basin

Hercynian structures

N

0   50 km

Figure 6.22   Hercynian fold structures in southwestern England and related granite intrusions.

a dolerite sheet was introduced into the northern Pennines to form the Great Whin Sill, which runs across Northumberland to the Farne Islands.

Throughout southern and eastern England (Fig. 6.18) Hercynian rocks are not seen at the surface but are known to extend beneath the Mesozoic cover. The Coal Measures are recorded in concealed coalfields in many places, including the eastward extensions of the Yorkshire, Nottinghamshire and Derbyshire coalfields, and in the Kent coalfield.

6.5.4   Alpine cycle

The Permo-Triassic period, during which the Alpine cycle started, was, like the Devonian, a time primarily of denudation. The Hercynian and relic Caledonian mountains were subject to erosion under desert conditions. Alluvial fans accumulated round the mountain fringes, and wind-blown sandstones were deposited in the lowlands of the south, the Midlands and on either side of the Pennines. Periodically, playa lakes existed, within which evaporite salts formed and fine mudstones were deposited. For a time the area east of the Pennines was occupied by an arm of the sea, within which the Magnesian Limestone was laid, as well as gypsum and rock salt.

At the end of this 'New Red Sandstone' deposition, a rise in sea level created continental shelf conditions across the south and east of the country as it had done in the Carboniferous. Throughout the succeeding Jurassic and Cretaceous periods, shallow water covered this area, and a sequence of separate clay phases (the Lias Clay, Oxford Clay, Kimmeridge Clay and Gault Clay) is divided up by limestones of even shallower-water phases. Many of these take the form of **oolites** (the Inferior and Great Oolite, Corallian and Portland Limestones), but the best known is the organic limestone that we call **chalk**. Occasional deltaic incursions into the sea produced the sandstones that occur in the sequence (e.g. the Wealden Sands and Greensands of Cretaceous age, and the Bagshot Sands of Eocene age). All these deposits were laid on a continental margin a long way to the north of the deep ocean basin within which eugeoclinal sediments were accumulating (i.e. the Tethys Sea).

The end of the cycle, the Alpine orogeny, followed the collision between Africa and Europe that created the Alps themselves (see Section 6.3). In Britain the effects of the orogeny were felt as the ripples created by far-off waves. The continental shelf deposits of southern Britain and adjacent parts of Europe were gently folded. In southern England the main push was from the south, and a series of folds with east–west axes was created. The synclines formed the London and Hampshire Basins (Fig.

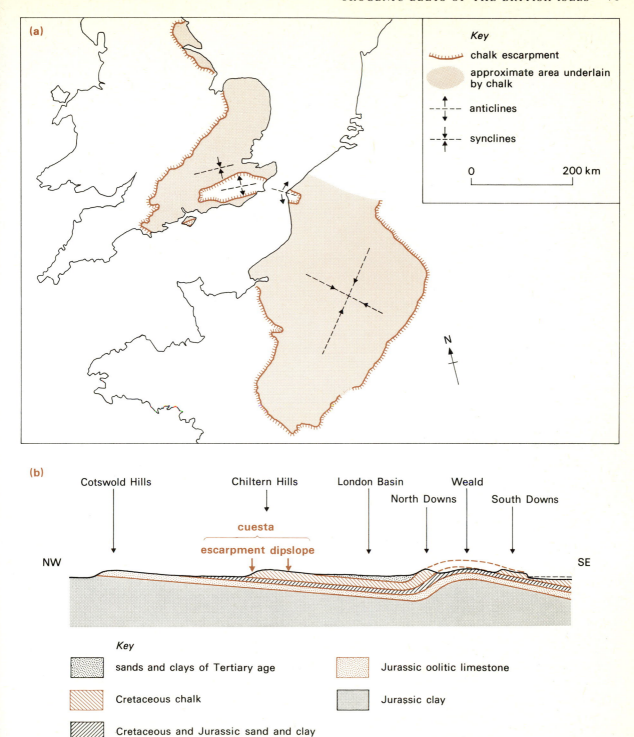

Figure 6.23 Alpine folding in southeastern England: (a) map showing the Mesozoic sediments of southeastern England as part of a wider synclinal basin; (b) section across the Mesozoic scarplands of southern England.

6.18) and the anticlines the Weald of Kent and the hills running through the Isle of Wight and the Isle of Purbeck. Only the most southern of these folds shows any degree of intense folding; strata along the Dorset coast are vertical in places. On a broader scale, however, the whole of southeastern England is really part of a large synclinal basin that extends into northern France (Fig. 6.23a), which gives an additional southeastward dip to the English rocks.

Subsequent denudation of the Mesozoic deposits in England has removed the less resistant clays and mudstones, with the result that the limestones and chalk tend to form upstanding ridges. The oolites form the Cotswold Hills, Northampton Heights, Lincoln Edge and North Yorkshire Moors ridge (although the northern section is deltaic sands rather than oolite). The chalk forms a ridge running from the Dorset coast, through Salisbury Plain and the Berkshire Downs, to the Chilterns, Lincolnshire and Yorkshire Wolds, with a subordinate ridge system forming the North and South Downs on either side of the Weald. Because the strata are tilted, each of these ridges is actually a **cuesta** (or tilted block), with a steep **escarpment** face and gentler **dipslope**. A section drawn across southeastern England (Fig. 6.23b) therefore shows a pattern of successive escarpments with intervening clay vales.

6.5.5 *Post-Alpine changes*

Even as the Alpine orogeny was reaching its peak in the area to the south of the British Isles, to the west the opening of the North Atlantic was progressing. The most recent splitting up of Laurasia had started well back in Jurassic times (Fig. 2.6b), since when the ocean had been expanding northwards. By Tertiary times (Fig. 2.6c) a spreading ridge had developed to the latitude of the British Isles and was in the process of dividing North America, Greenland and Europe. The edge of the rift system must have passed very close to the northwestern corner of the British Isles, because volcanic activity (Section 4.3) suddenly commenced in northwestern Scotland and Northern Ireland during the early Tertiary. Today, the centres of the activity can be seen as eroded vent complexes on some of the Inner Hebrides islands and in the Mourne Mountains (Fig. 6.24). These complexes are surrounded by swarms of dykes intruded into the country rocks and radiating out from the vents. More important are the extensive basalt-lava plateaux that cover the Antrim area of Northern Ireland and large parts of Skye, Mull and Lorne. At the coast the lavas form

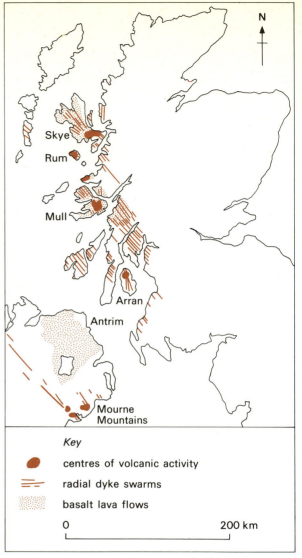

Figure 6.24 Tertiary volcanic activity in northwestern Scotland and northern Ireland.

cliffs, and the peculiar hexagonal columns produced by cooling of the basalt are notable features of the Giant's Causeway at the edge of the Antrim Plateau and on the island of Staffa.

Since the Atlantic split apart, the British Isles have been on the trailing edge of the European continental areas as it travels eastwards. The continental shelf and slope around Britain have therefore been a continuing site of marine deposition, such as the corresponding American seaboard (Ch.

5). Most of the land area of these islands has remained a denudation zone since the mid-Tertiary, providing sediment to the surrounding seas. During the Pleistocene much of the country was covered by ice sheets on a number of occasions, and this has left a veneer of glacial moraine, which may be locally quite thick. Otherwise, deposition has been largely restricted to the coastal zone, although the sea level changes associated with the glaciations has led to the silting of a number of estuaries, producing the marsh areas of the Fens, Somerset Levels and other major rivers.

The British Isles have therefore been close to tectonic activity on a number of occasions since the Precambrian remnant of northwestern Scotland was formed. The gradual addition of sections of orogenic belt of varying intensity and lithology has done much to give Britain the enormous diversity of scenery that is visible today.

Chapter 7

Volcanoes and Earthquakes

Throughout the preceding chapters we have made frequent reference to both volcanic and earthquake activity as an indication of, and a consequence of, the movement of the lithosphere. A further chapter is devoted to these phenomena, in order that we may examine the characteristics of volcanic eruptions and landforms and the impact of volcanoes and earthquakes upon man's activities.

7.1 The distribution of volcanoes and lava types

The world distribution of volcanoes was originally described in Figure 3.10 as an indication of zones where heat flow from the Earth's interior is higher than average. Subsequently, this information was used to define the tectonically active areas of the world and therefore the margins of lithospheric plates. We can now look at the world distribution of volcanoes in tectonic terms. On this basis active volcanoes may be found in the following locations:

(a) Along the spreading oceanic ridges where up-welling mantle material creates lithosphere on a constructive plate margin. The Atlantic ridge seems to be richer in volcanoes than the spreading ridges of the Indian and Pacific Oceans, but this may be simply a result of the failure to record submarine eruptions in remote locations.

(b) Along continental rift valleys which may represent incipient or failed continental splitting. The East African rift valley system is the only existing location of this type.

(c) In island arcs created by the subduction of oceanic lithosphere. The vast majority of these arcs lie in the great belt stretching from the Aleutian Islands in the north to the Kermadec Islands in the south, which marks the subduction of the Pacific plate. Other arcs include Indonesia, the West Indies and the Scotia Ridge in the South Atlantic.

(d) In recent orogenic belts where at least one continental area is in collision with a subduction zone. This includes the single-continent collision belts of the Andes and western ranges of the North American mountains as well as the double-continent collision belt of the Alps.

(e) At isolated locations inside the boundaries of a lithospheric plate, where 'hot spots' are created by rising plumes of material within the mantle. Most of these are found within the borders of the Pacific plate.

As far as the volcanoes themselves are concerned, these differing tectonic environments fall into two categories by lava source and type:

(a) The magma involved in spreading ridge, rift valley and hot spot volcanoes is produced by the upward movement of mantle material. As such, the lavas produced by this magma reflect the chemistry of the mantle and are therefore broadly of **basaltic** composition (Appendix A.2).

(b) The magma found in subduction zones, whether island arc or orogenic belt, has its origin in the partial melting of some of the lower temperature and less dense minerals from the subducting oceanic crust. The subsequent lavas are therefore broadly of **andesitic** composition (Appendix A.2).

The geographical distribution of volcanoes therefore also has a geochemical significance. From our point of view, however, those geochemical differences have enormous importance, because they have a profound effect upon the nature of a volcanic eruption and ultimately upon the landforms produced by the eruption.

7.2 Lava type, eruption pattern and volcanic products

Lava (i.e. partially melted rock at the surface) is the most obvious, although by no means the most important, product of a volcanic eruption. The main types of lava (e.g. basalt, andesite and rhyolite, Appendix A.2) are distinguished on the basis of their mineral composition, but that in turn has an important bearing upon the **viscosity**, or flowing properties, of the lava. Basalt, for example, is normally very free flowing, with something like the consistency of syrup; it forms bubbling lava lakes or fast-moving rivers of molten rock. Andesite, on the other hand, is much more viscous and is often described as 'pasty'; it moves only slowly and tends to produce lava flows that are thicker but less extensive than basalt flows. In some cases the lava does not move as a true flow but creeps forwards as a pile of lava blocks. Rhyolite is often so viscous that it does not move away from the vent of a volcano but is forced up into dome-like mounds called **tholoids**, from which spines of hot, but virtually solid, lava are squeezed vertically.

The reasons for this range in viscosity are fairly complicated. To some extent temperature has an influence. Basalt is composed largely of minerals that become liquid only at high temperatures, with the result that flowing basalt is usually around 1100 °C. Rhyolite, in contrast, consists of minerals with a lower melting temperature, and the rock may be in a semi-liquid state at only 650–700 °C. Since viscosity decreases with temperature, basalt is less viscous than rhyolite. On the other hand, the actual chemistry of the lava seems to have some effect also, since andesite remains more viscous than basalt, even though often found at a higher temperature.

Lava viscosity is clearly important in influencing the shape of landforms built by lava, but of greater significance is the influence of lava viscosity upon the nature of the eruption itself. The magma from which lava is derived also includes large volumes of dissolved gases (e.g. water vapour, carbon monoxide, carbon dioxide, hydrogen sulphide, sulphur dioxide). Normally, the magma is confined under high pressure, which keeps the gas in a dissolved state; but if the pressure drops, the gas will exsolve from the magma with a consequent rapid expansion of volume. It is this rapid expansion of exsolving gas that actually forces magma up to the surface through the vent of a volcano – a process not entirely dissimilar to the effect produced when the top is removed from a bottle of fizzy drink after it has been shaken.

Many lavas when cooled are seen to be full of gas escape holes called **vesicles** – hence, **vesicular lava**. At this point, however, viscosity becomes important. A lava with a low viscosity (e.g. basalt) enables gas to escape freely, whereas a more viscous lava tends to trap gas until the pressure is sufficient actually to blow the lava apart. The eruption of andesite therefore is often accompanied by a good deal of explosive activity, during which lava is blown out of the vent in fragments. Fragmented lava of this type is usually called **pyroclastic** material (literally, 'fire-broken'). Most of the pyroclastic accumulation around a vent settles through the air, and this is known as **tephra**. Generally, the thickness of a tephra deposit decreases with distance from the vent, as does the average size of particles making up the tephra. Words like **bomb**, **lapilli** and **ash** are used to describe tephra of decreasing size. In many cases gas continues to be released within the tephra deposit. The term **scoria** is often used for the highly vesicular tephra blocks that accumulate close to a vent, and the familiar **pumice** is really fine-grained vesicular tephra.

Under certain conditions, usually involving highly viscous lava, the explosive force of exsolving gas is not directed upwards into the atmosphere but instead pushes outwards and down the sides of the volcano. This is somewhat akin to the ground-level shock wave that moves out from the point of an atomic explosion. In this instance the pyroclastic material and gas are thoroughly mixed in a very hot cloud, which may roll away from the vent at great speed – a phenomenon known as a **nuée ardente** (literally, 'glowing cloud'). The pyroclastic material carried by the nuée is laid under conditions similar to those occurring in a fluid medium, and it is preferable to describe such deposits as **pyroclastic flow deposits** rather than tephra, which is produced by pyroclastic airfall. Very thick deposits called **ignimbrites** are often found in older geological formations, and these may have their origin in nuée ardente eruptions.

The variation in lava chemistry represented by the basalt–andesite–rhyolite sequence therefore causes differences in lava viscosity, which partly affects the form of lava flows and partly affects the volume and nature of pyroclastic material. Thus, basalt has a low viscosity and tends to give rise to eruptions where large volumes of lava flows occur but there is little pyroclastic material. In contrast, a typical andesite eruption produces less lava, which is generally more inclined to pile up at the vent, plus a much larger volume of tephra. At the extremely viscous end of the range, an eruption may produce virtually no recognisable lava, all the products of eruption being in pyroclastic form. Furthermore, volcanoes with a more viscous lava tend to erupt only sporadically, since the vent can easily become blocked. Consequently, whereas many basalt volcanoes are in a more or less constant state of moderate activity with frequent lava outpourings, many andesite and rhyolite volcanoes may lay dormant for years, or even be considered extinct, before erupting with great violence at a point where the buildup of gas pressure is sufficient to overcome the strength of the old lava blocking the vent. In volcanoes where periodic violent eruptions occur, the shape of the volcano may be determined as much by explosions as by the accumulation of lava or pyroclastic material.

7.3   Examples of the main types of volcanic eruption

In one sense every volcano is unique, and that is partly what makes them so fascinating. On the other hand, it is possible to identify typical eruption patterns and volcanic landforms that apply to a whole group of volcanoes. In this section we shall examine some of the eruption types recognised by vulcanologists. To some extent, these eruption types are based either upon the nature of the lava involved or upon the tectonic environment, but the reader is warned against making too simple a connection in this respect.

7.3.1   Icelandic eruption

The volcanic eruptions experienced regularly in many parts of Iceland are typical of those which occur along spreading oceanic ridges and continental rift valleys during the splitting of continental crust. In most cases, the eruption centres do not resemble the normal picture of a volcano. There is no single cone with a central vent, for example; instead, lava issues from numerous points along a **fissure** in the Earth's surface (Fig. 7.1).

The largest Icelandic eruption in recorded history took place in 1783. The Laki volcano started to erupt at no fewer than twenty-two separate vents along a 25 km fissure. Initially, the eruption produced huge volumes of sulphurous gas and ash, which was scattered over a wide area downwind of the fissure. Small scoria cones developed around each vent. The main feature of the eruption, however, was the outpouring of a huge volume (approximately 11 km³) of basalt lava, which rapidly spread into the adjacent Skaftar river valley, eventually producing lava flows that were 56 km in total length. Lavas flowing freely in this manner from the length of a fissure tend to fill up valleys and hollows in the landscape until a flat plateau is produced. Repeated eruptions from fissures at different locations can result in enormous thicknesses of more or less horizontal basalt, and it is not always possible to locate the fissure itself. It is generally agreed that the flood basalts that compose the world's great basalt plateaux (Section 4.3) were erupted in this manner.

In the 1783 Laki eruption, although the lava flows destroyed some agricultural land, they did not in themselves pose any serious threat to life. Nonetheless, it is estimated that 10 000 people, who represented one-fifth of the population of Iceland at the time, died as the result of the eruption. This terrible loss of life was due primarily to the effects of gas and ash upon crops and livestock, which were so devastating that famine occurred on a huge scale.

While looking at the Icelandic situation, it is worth noting that this is one of the few locations in the world where subglacial eruptions can occur. In 1934, the Grimsvötn volcano erupted beneath the Vatnajökull ice cap (Fig. 7.1). The effect was to melt the ice from the inside. At first little water escaped; but once melting had reached the edge of the ice cap, the ponded water inside was suddenly released. It is estimated that something like 100 000 m³/s flowed from the Vatnajökull – a figure that represents half of the normal discharge of the Amazon! The destruction can be imagined.

Both the 1783 Laki and 1934 Grimsvötn eruptions serve to make the point that it is rarely the flow of lava that causes the damage during a volcanic eruption. The same picture emerges when we examine other eruption types.

7.3.2   Hawaiian eruption

The Hawaiian Islands are formed by the peaks of a line of basalt volcanoes, which become progressively older to the northwest. Their formation is interpreted as being due to the passage of the Pacific plate across a stationary 'hot spot' (Section 5.1). Hawaii is now the only island with active volcanoes. Of its three peaks, Mauna Kea is extinct, leaving Mauna Loa and Kilauea as the active vents (Fig.

7.3). The island as a whole is constructed of countless overlapping basalt flows piled up around the central vents. In cross-section (Fig. 7.2) it can be seen that each volcano consists of a very high (up to 10 000 m above the ocean floor) but low-angle basalt dome called a **shield volcano**.

At the summit of Mauna Loa is a crater about 5 km long and 2.5 km wide. This crater has been created largely by collapse of the volcano summit into the underlying magma chamber and is referred to as a **caldera**. Eruptions, which occur on average about every 3.5 years, usually start with activity in this summit caldera. In January 1949, for example, the 200 m deep caldera filled with molten basalt and became the centre of 'fire fountain' activity, as gas bubbling up through the lava flung sprays of liquid material high into the air. Eventually, the fountain activity began to build a scoria cone within the caldera, which grew until it rose above the caldera rim. After some time, the level of lava topped the caldera rim on the southern side, and a stream of basalt poured out to produce a long but narrow flow. Beneath the volcano the magma level remained high, however, because in June of the following year (1950) a small earthquake was followed by the opening of a fissure 4 km long along

the crest of the ridge that stretches southwestwards from the caldera (Fig. 7.3), from which lava began to pour. Fissures continued to open up along the southwestern ridge as this **flank eruption** gathered momentum, and ever larger volumes of basalt were ejected in streams that eventually ran down into the sea.

The lava streams produced during an Hawaiian eruption may exhibit numerous different forms. The base and sides of a flow often solidify quickly by contact with cooler surfaces, so that fresh lava seems to run between walls of rubbly vesicular basalt. These walls are similar in appearance to the levées along a river bank, after which they are named. As the surface of the lava chills, it may retain the shapes of the flowing liquid, in which case it is called 'ropey' lava, or **pahoehoe** in Polynesian. Lava flowing beneath a cooling surface may cause the surface to break up into angular pieces, which are given the name **aa** lava. It is quite possible for the lava surface to become completely solid while liquid material continues to flow beneath the surface in tunnel networks.

The Hawaiian eruption is thus typified by fairly regular outpourings of liquid basalt from a central vent or flank fissure, which accumulate to form a

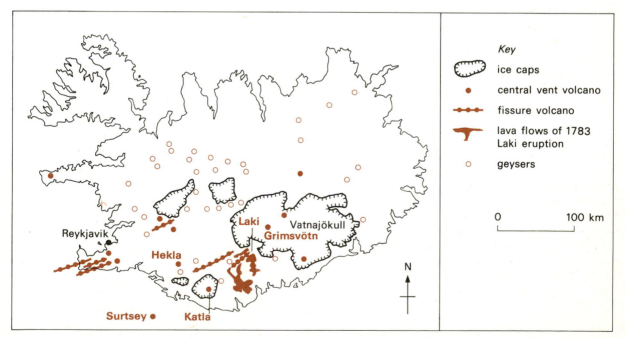

Figure 7.1   The location of volcanoes in Iceland. This map also shows the extent of lava flows from the 1783 Laki eruption.

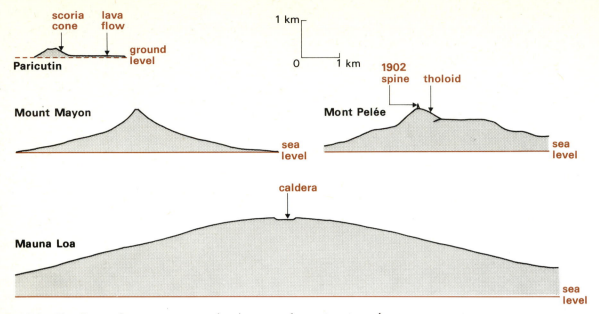

Figure 7.2 The shape of some common volcanic cones, drawn true to scale.

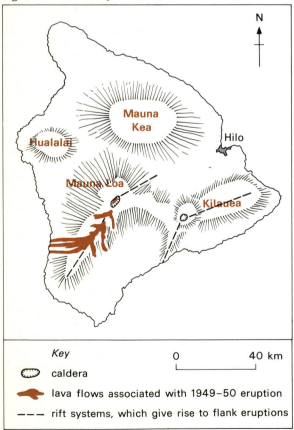

Figure 7.3 Hawaii: the Mauna Loa volcano and the lava flows of the 1949–50 eruption (after Bullard 1962).

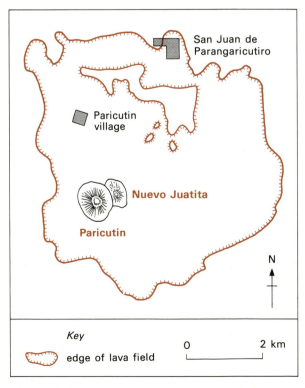

Figure 7.4 Paricutin volcano, Mexico, and the cone and lava field in 1952 (after Bullard 1962).

shield volcano. The amount of pyroclastic material produced is very small, and the active phases of the cycle are fairly quiet. Loss of life is rare, and damage to property is mainly confined to land and buildings actually overwhelmed by flowing basalt.

7.3.3 Strombolian eruption

Stromboli is an island lying between Sicily and mainland Italy and therefore falls within the Alpine collision belt (Fig. 6.12). Although a basalt volcano, its lava is somewhat less fluid than the basalt of Hawaii, with the result that gas escapes only spasmodically, producing small explosions. In each explosion semi-liquid lava is hurled into the air to form a **spatter cone** of scoria and bomb material around the vent. It is the construction of this scoria cone by almost continuous activity that is the distinctive feature of the Strombolian eruption. In the case of Stromboli itself, the cone rises with almost perfect symmetry to a height of 930 m above the ocean floor. Because the cone is produced by falling scoria rather than by flowing lava, the sides of Stromboli are much steeper (about 33°) than the sides of Mauna Loa in Hawaii (about 12°). Long years of relatively quiet Strombolian activity may be interrupted by occasional more violent episodes, during which ash and lava are released.

Even as these lines are being written (August 1979) news is coming in of an eruption by Mount Etna, which normally displays Strombolian activity. Fortunately, the eruption is not likely to cause any loss of life, but lava has destroyed 50 hectares of orchard, and the fall of ash is causing considerable inconvenience to the tourist industry.

The eruption of the Paricutin volcano in Mexico in 1943 not only illustrates the growth of a typical Strombolian cone but also serves to demonstrate that any volcano has to start at some point in time. At 4 pm on 20 February 1943, Dionisio Pulido was walking through one of his fields when he noticed a small crack, about half a metre deep, in an already existing hollow. As he watched, the ground around the fissure began to swell, and sulphur gas and fine ash started to drift out of the crack. Within minutes small globules of molten material like sparks were being ejected, and some of the nearby trees started to burn. By 5 pm the gradually growing column of ash and smoke could be seen from the village of Parangaricutiro 5 km away, and substantial amounts of scoria were being thrown out of the hole. The scoria mound attained a height of 10 m within 24 hours and kept growing. On 22 February, black jagged lava emerged from the northeastern

corner of the cone and spread out, eventually overwhelming Pulido's farm. During the whole of 1943 the activity continued. Sometimes the lava flows diminished and the cloud of ash seemed to increase, but at other times the lava poured out with renewed vigour. Throughout this time the scoria cone kept growing; within one week it was over 100 m high, and after a year it had risen to 310 m. After that point the growth of the cone slowed, but lava continued to flow during 1944, eventually covering the villages of Paricutin and Parangaricutiro (Fig. 7.4). After the end of 1944 all forms of activity decreased, and subsequent lava flows were superimposed upon earlier flows or filled in gaps to produce an irregular lava plain. A temporary increase in activity in 1946 caused the opening of a second vent (Nuevo Juatita), which added another scoria cone to the first. Eruption finally ceased in 1952, when the main scoria cone was 400 m high.

Strombolian activity seems to occur at subduction zones or collision zones where at least one continent is involved. The lava seems to be similar to basalt, but it may have slightly unusual properties as the result of chemical interaction between the rising magma and overlying continental material.

7.3.4 Vesuvian eruption

Vesuvius is probably the most famous volcano in the world. It stands a little way to the southeast of Naples on the western coast of Italy and has given its name to some of the more spectacular aspects of volcanic eruption. The main feature of a Vesuvian eruption is a longish period of inactivity during which gas pressure builds up behind the lavas that clog the vent. The blockage is removed by a substantial explosion, or series of explosions, during which large volumes of pyroclastic material (especially ash) are blown out of the vent by a sustained blast of gas. In the later stages of eruption, relatively small volumes of lava, usually andesitic, are released to form thick blocky flows, which do not spread over any great area.

The classic eruption by Vesuvius itself occurred in AD 79. A series of minor earthquakes preceded the eruption but seems to have been ignored. The eruption itself started on 24 August and lasted for two days in the main phase. During that time a series of massive explosions sent a huge cloud of ash into the air, effectively concealing the volcano. Observers (e.g. Caius Plinius) who attempted to approach the volcano by sea found it impossible to land, as the shoreline was completely clogged with ash and layers of pumice, which floated on the surface of the bay. At the end of the eruption, when the ash cloud had dispersed, it was clear that enormous changes had occurred to the shape of the volcano (Fig. 7.5). The removal of the summit may have been partly explosive, but probably much of the change in shape occurred as the top of the cone collapsed into the emptying magma chamber beneath. Certainly, an incredible quantity of ash had been ejected. The town of Pompeii was buried beneath 3 m of ash and pumice, and substantial ashfalls were observed a long way downwind (Fig. 7.6).

The AD 79 eruption is particularly well known because it resulted in the remarkable preservation of the towns of Pompeii and Herculaneum. Pompeii, it has already been said, was steadily overwhelmed by airfall ash and pumice during the two days of the eruption. Buried within the ash were the bodies of some of the inhabitants. On the whole it seems that most people escaped from the town and that those who died were overcome by fumes before being buried by ash. Herculaneum, on the other hand, was upwind of the volcano (Fig. 7.6) and only received a light ashfall. However, it was buried, probably within a matter of minutes, by an enormous mudflow created by the action of torrential rain (caused by water vapour and ash released during the eruption) upon the piles of ash falling on the upper slopes of the volcano.

The AD 79 eruption was a particularly violent episode in the history of Vesuvius. In recent centuries the volcano has erupted on average about

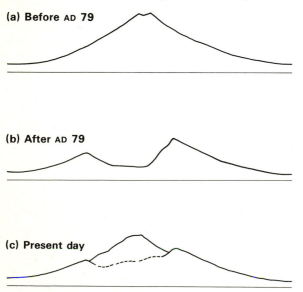

(a) Before AD 79

(b) After AD 79

(c) Present day

Figure 7.5 The changing outline of Vesuvius, Italy.

once every forty years, with the release of a substantial amount of ash and a small volume of lava. Where volcanoes are subject only to this milder form of Vesuvian activity, a near-perfect symmetrical cone is produced, which consists largely of tephra layers plus minor lava flows. A **composite cone** of this type is often regarded as the 'normal' shape of a volcano. Fujiyama in Japan has a very fine shape, although Mount Mayon in the Philippines (Fig. 7.2) is actually more symmetrical. Vesuvius, however, is far from a simple cone, because the long periods of moderate activity are interrupted by occasional eruptions of the AD 79 variety, which are sometimes called **Plinian eruptions**. Plinian eruptions seem to cause the collapse of the cone on Vesuvius, and secondary cones are subsequently built within the collapsed feature (Fig. 7.5). The outline of modern Vesuvius is therefore due to both constructive and destructive processes.

Another volcano experiencing Vesuvian-like eruptions, where the destructive element is even more important, is the island of Krakatoa, which lies in the Sunda Straits between Java and Sumatra. Krakatoa erupted in 1883 and before that date was simply known as an uninhabited island clothed in dense jungle. Explorers had observed a series of peaks running along the centre of the island and increasing in height to the south. They had been identified as having a volcanic origin and were named, in order of height, Rakata, Danan and Perboewatan (Fig. 7.7). Although no eruption had been recorded previously, there was a recent lava flow on the side of Perboewatan.

The story started on 20 May 1883 with a series of loud explosions, heard up to 150 km away, and the ejection of an ash cloud from the Perboewatan crater some 11 km into the atmosphere. Allowing for occasional lulls, the explosions and steady outpouring of ash continued for the next two months. During June it was reported that Danan was also in eruption, and by August Rakata itself had started, so ash and gas were escaping from three separate vents. Nonetheless, this prolonged activity was nothing more than the prelude to the main eruption, which took place during 26 and 27 August. On these two days the whole area shook to the sound of a series of explosions so huge that the previous noises palled into insignificance. The crescendo was reached at 10.02 am on the 27th, when an explosion occurred that was heard on the island of Rodriquez, near Mauritius in the Indian Ocean, no less than 4811 km away. The noise was loud enough to wake people at Elsey Creek in South Australia,

3224 km away, and the shock wave blew out windows in Batavia, the capital of Java, some 160 km distant.

The explosions were accompanied by two other events. First, as in the Vesuvius eruption of AD 79, an enormous quantity of fresh ash and pumice was ejected. At the height of the eruption the ash cloud was estimated to have risen 80 km into the atmosphere. For two days the sun was blotted out entirely in the area around the Sunda Straits, and there was a substantial ashfall in Batavia. The very finest dust reached so far up into the atmosphere that it seems to have become trapped in the upper air jetstreams in such a way that it rapidly circled the Earth and gave rise to spectacular sunsets in many parts of the world for months after. Eventually, 1 cm of dust from the volcano settled on to Madrid! The second event also paralleled the Vesuvius eruption. As ash was being blasted out of the volcano, the magma chamber emptied until collapse of the cone occurred. Unlike Vesuvius, however, the collapse of the island set up shock waves in the water, which spread out from Krakatoa as a massive 'tidal wave' (or, more properly, **tsunami**) 40 m high. The tsunami overwhelmed the coastal settlements around the Sunda Straits, causing the deaths of 36 000 people.

After the activity had ceased, exploration parties were sent out, which found that the island had virtually ceased to exist. The formation of a caldera

by collapse had turned the 300 m high Perboewatan cone into a 300 m deep lagoon (Fig. 7.7).

7.3.5 Peléean eruption

Like Krakatoa, Mont Pelée is located in an island arc. In this case the volcano is found on the island of Martinique, which is part of the group called the Lesser Antilles lying between the South American mainland and the Caribbean island of Puerto Rico.

Mont Pelée lies at the northern end of Martinique (Fig. 7.8) and rises to a height of 2250 m. A small caldera at the summit was once occupied by water, but at the end of the nineteenth century this was dry and so was called the Étang Sec. The town of St Pierre, capital of Martinique and a thriving port of some 20 000 inhabitants in 1902, lay 7 km southwest of the summit. Pelée had erupted in a mild way within living memory, and the start of the activity in 1902 came as no surprise.

The early stages were unremarkable and included the start of steam and gas release from the summit, accompanied by mild tremors. During April 1902 the Étang Sec filled with hot water, through which a small ash cone projected. Light falls of ash and the presence of gas were reported in St Pierre. On 5 May the water in the Étang Sec seems to have overflowed and rushed down the valley of the River Blanche, collecting ash and mud as it moved until it took the form of a boiling mudflow. This mudflow destroyed

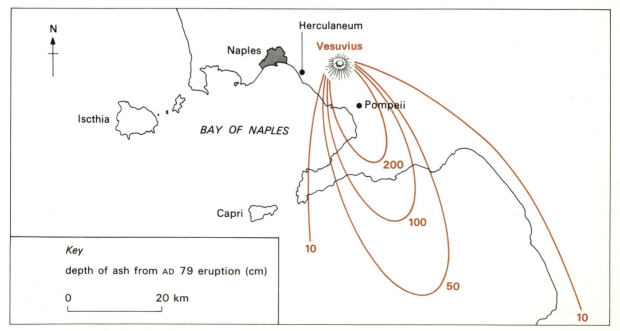

Figure 7.6 Isopach map showing the depth of ash fall from the AD 79 eruption of Vesuvius (after Francis 1976).

a sugar mill in the lower part of the valley, killing the workers, and sped into the waters of the bay, causing some minor tsunami, which capsized a few small vessels. The population became alarmed by these events but was dissuaded from leaving the area as a whole, although there was a movement from the surrounding countryside into St Pierre. Within a few days the population had swollen to 30 000.

At about 7.50 am on 8 May, four loud explosions occurred in quick succession, and a black cloud was seen to emerge from the summit. Instead of simply rising upwards, however, the cloud started to flow down the sides of the volcano in the direction of St Pierre. This was the first recorded instance of a nuée ardente. Sweeping down the slope of the mountain

at a speed in excess of 150 km/h, the nuée passed through the town and into the bay within a matter of minutes. After it had passed, the town was a mangled wreck of burning debris, the bay was littered with the charred remains of boats, and 30 000 people were dead. Picking over the ruins after the event, it became clear that damage to property had been caused initially by the sheer force of the blast of the gas cloud, since many masonry walls had been demolished. The extremely high temperature of the gas and the particles of ash that it carried had then set the combustible material on fire. The town continued to smoulder for a full week after the event. In nearly every case death had been due to the inhalation of superheated gas and had been instantaneous; people had literally dropped where they stood. The number of survivors was incredibly small. Out of the entire population of the town, only four (two in some accounts) were left alive, and in each case the individual had been partly protected by being in a closed room at the

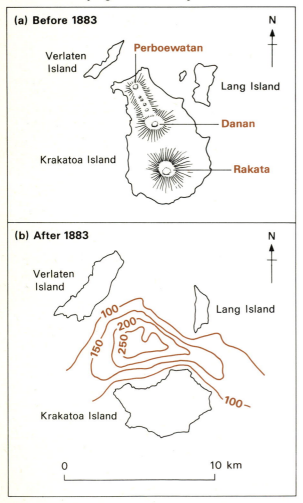

Figure 7.7 Before and after maps of Krakatoa, Indonesia, for the 1883 eruption. The contours show water depth in metres (after Bullard 1962).

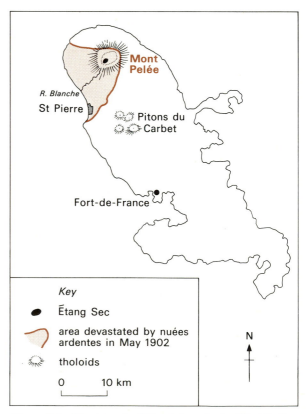

Figure 7.8 The island of Martinique in the West Indies, showing the area of devastation caused by the nuées ardentes of the 1902 eruption of Mont Pelée (partly after Francis 1976).

time of the disaster. The most famous survivor was a prisoner in the town gaol.

Over the next few weeks a number of nuées ardentes flowed over the same area but could do no more damage. Meanwhile, ash continued to fall continuously in the ensuing months. On 30 August the last destructive episode occurred, when fresh nuées were ejected but flowed off in a new direction, destroying one village and killing a further 2000 people. Even then the eruption was not finished, since in October a tholoid dome (Fig. 7.2) began to rise on the site of the Étang Sec. Cracking must have occurred in the surface of the tholoid, since a lava spine of **dacite** composition (a quartz-rich variety of andesite) began to be squeezed out from one side. The spine had a diameter of 150 m and grew steadily in height, despite the fact that it was constantly subject to disintegration. The spine reached a maximum height of about 340 m in May 1903, after which the disintegration rate exceeded the rate of growth.

The 'classic' eruption patterns described above tend to reinforce the idea that basaltic lavas flow freely and produce frequent eruptions with large volumes of lava and limited explosions, whereas andesitic lavas erupt less frequently but with greater violence and a much greater production of pyroclastic material. It should be borne in mind, however, that individual volcanoes may produce different lavas and different eruption patterns at various points in their history.

The descriptions also perhaps serve to illustrate that loss of life is rarely caused by the eruption of lava. Normally, there is plenty of time to move from the path of a lava flow, and in recent years it has not been unknown to try diverting lava flows by bombing. In the historic eruptions that caused great loss of life, death was caused by some other related phenomenon (e.g. gas, mudflow, tsunami). In many cases also the warning signs in the form of pre-liminary tremors or explosions went unheeded. Today, we understand volcanoes rather better, and a close watch is kept on most that are close to centres of population. In developed countries at least, future disasters should be minimised by the rapid evacuation of population from an area once renewed activity has been identified.

7.4   Earthquakes in a world setting

The immediate cause of an earthquake was identi-fied in Section 3.2 as the elastic rebound of rock

material back to its original shape after failure has released the stress causing deformation (Fig. 3.1). In most cases the stress is produced by the opposing motion of moving bodies of rock, and repeated failure along a single line produces a fault. In Figure 3.11 it was observed that the world distribution of earthquakes follows the same linear pattern as the distribution of volcanoes, and this information helped to define the limits of the rigid lithospheric plates. Having examined the lithospheric plates and their boundaries, we are now in a position to summarise the main tectonic environments that produce earthquakes.

Minor earthquakes can occur almost anywhere in the world. These may be the result of movement along old faults, associated with isostatic changes, or quite commonly, they may be due to changes in lava level in the magma chamber beneath a volcano. Nonetheless, the overwhelming majority of the more intense earthquakes occur along the belts identified in Figure 3.11.

A small but significant number of earthquakes is recorded from along the spreading oceanic ridges. Doubtless, many more go unrecorded, since they have little surface effect. The main locations of earthquake foci on the spreading ridges are along the vertical faults that bound the central rift and along the transform faults that offset the ridge. The more general fracture zones extending beyond the ridge crest are less likely to generate earthquakes, since the surface is moving in a common direction. In all these locations the focus is likely to be quite shallow.

A much greater concentration of earthquakes can be found along the subduction and collision zones of the world. The subduction zones in particular repeatedly show evidence of a Benioff zone extend-ing from the surface to a maximum depth of 700 km in the direction of movement of the subducting plate. Benioff zones are recognisable beneath both island arcs and marginal orogenic belts (e.g. the Andes). Although collision zones (e.g. the Alps, the Himalayas) are seismically very active, earthquake foci tend to be shallow, probably because there is no longer a subducting plate beneath the mountains.

7.5   The effects of earthquakes and man's response to the hazard

The seismic waves that penetrate the interior of the Earth and are of so much help in unravelling the internal structure of the planet are set in motion by earthquakes but do not have any marked effect upon the surface. On the other hand, the same earthquake can also set in motion a series of surface waves, usually called **long waves** or **L-waves**, which radiate out from the epicentre. It is the movement of the ground in response to the passage of these surface L-waves that causes the damage during an earthquake. The actual pattern of ground motion is very complicated and includes vertical as well as lateral components. The motion takes the form of sudden accelerations backwards and forwards along each of the component directions. It is the rate of acceleration, rather than the actual amount of movement, that is critical in causing the collapse of structures. The intensity, or magnitude, of an earthquake is measured on the **Richter scale**, which takes as its value the logarithm of the maximum amplitude recorded on a seismograph trace at a set distance from the epicentre. A Richter value of two can only just be felt, at five local damage occurs, and anything over seven is regarded as a major earthquake.

Major earthquakes are much the same wherever they occur, unlike volcanoes. On the whole the main natural effect is to loosen material, creating movements on slopes (e.g. landslides, rockfalls, avalanches). The extent of such movement tends to increase with the earthquake magnitude and is obviously greater in mountainous areas, where slopes are already steep. The effect that an earthquake has upon man also varies with the same factors, and the damage must clearly be more extensive in more heavily populated areas. A major factor that plays a part particularly in determining the loss of life in an earthquake, however, is the socio-technological level of the society affected. A few examples should clarify this point.

7.5.1 Earthquakes on the San Andreas Fault zone

Southern California lies astride one of the most active earthquake zones in the world. The tectonic setting is far from simple, but we can probably regard this as part of a spreading ridge system. There appears to be a spreading ridge extending into the Gulf of California (Fig. 4.15), which is connected to another ridge element off the Oregon coast (Fig. 4.4) by a series of transform faults. These transform faults run across southern California, giving an intricate pattern of surface faults (Fig. 7.9). The master fault in this belt has long been known as the San Andreas Fault. Movements along the San Andreas appear to be causing Baja California and the coastal strip around Los Angeles

to move northwestwards relative to the rest of California. It is this transform movement that is primarily responsible for earthquakes.

The most famous of the San Andreas movements occurred at 5.12 am on 18 April 1906. The resulting earthquake had a magnitude of 8.3 on the Richter scale. The San Andreas seems to have opened up over a distance of nearly 300 km, from Point Arena to San Juan Bautista, along which there was a lateral movement of 5–7 m. The shock waves caused major structural damage to San Francisco and in coastal areas along the length of the fault, as well as in sedimentary cover areas within the Central Valley (Fig. 7.10). In San Francisco itself, however, even more destruction was caused by the fires that followed the earthquake. The fires were due in large part to the fracturing of gas mains, while the corresponding fracture of water mains made the fires difficult to extinguish. By the end the city lay in ruins, although only 700 people had lost their lives.

In the years since 1906, considerable effort has been made to reduce the effects of earthquakes in this highly vulnerable urban–industrial system. Building codes have tried to make man-made structures safer. In some cases a height restriction is in force; some buildings are designed to 'rock' or have 'floating' foundations; others will collapse into a rigid box. Regular earthquake drills in schools and factories, and education of the population, particularly in after-earthquake precautions (e.g. avoiding contaminated food and water, keeping adequate stores in readiness), are all intended to minimise loss of life. Exactly how effective all these measures will actually be when put to the test is difficult to assess. On 9 February 1971, for example, an earthquake with a Richter magnitude of 6.6 hit the suburbs of San Fernando. Inevitably, many old buildings collapsed, but so did some built to the new regulations. Only sixty-four people were killed, but damage was considerable, perhaps as much as $1 billion worth. Potential disaster came very close when part of the Van Norman Dam collapsed, leading to the enforced evacuation of 80 000 residents of the area.

Despite the hazards, however, it is often difficult to persuade a whole population of the risks involved. In the years since 1906, San Francisco has been rebuilt as a thriving city. Its bustling central business district has grown upwards, until the skyline is dominated by tower blocks and the streets are full of people and cars. But one day fairly soon the San Andreas will move again. Because the forces acting along the San Andreas are more or less constant, it follows that on average a certain amount of movement must take place each year. Along the southern part of the fault this figure is 62 mm/yr. The longer the time interval between successive earthquakes, the greater will be the movement when it eventually occurs. At this point in time (summer 1979) no movement has taken place in the San Francisco area for some years. It is now generally expected that the next earthquake will be a 'big one', and there is a fifty–fifty chance of its happening in the next decade. A recent report has estimated that, if another Richter scale 8.3 earthquake struck San Francisco during the rush hour, the toll would be 10 000 dead and 50 000 injured.

7.5.2 Earthquakes in underdeveloped countries

Although the figures for predicted casualties in the event of a major San Andreas movement may appear frightening, they are nothing compared with estimated death tolls for earthquakes in other parts of the world. During this century major disasters have struck several times: for example, Kansu, China, in 1920 (180 000 dead); Tokyo in 1923 (143 000); Kansu again in 1932 (70 000); Quetta, Pakistan, in 1935 (30 000); and Chile in 1939

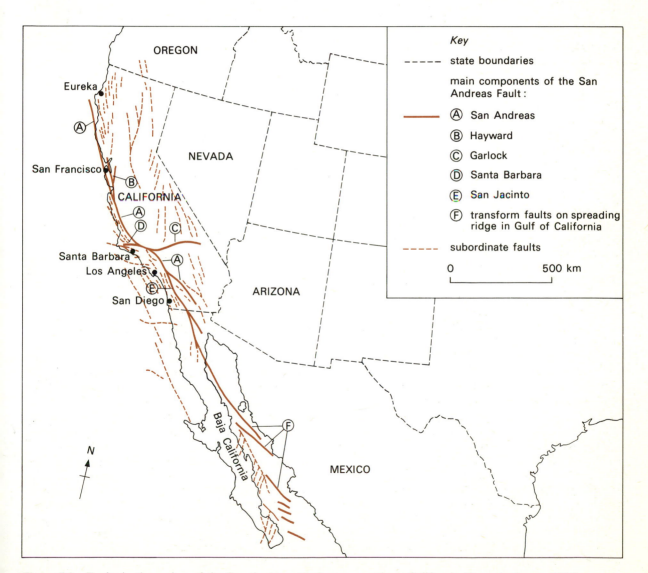

Figure 7.9 Faults in the region of the San Andreas Fault zone, southern California (from Anderson 1971).

(40 000). An estimate of the death toll from an earthquake in Shenshi province, China, in 1556 puts the figure as high as 830 000.

In nearly all these cases the story is the same. The deaths resulting directly from the earthquake, from being hit or trapped by the collapse of buildings, represent only a small proportion of the final count. Over and over again the huge death toll is caused by exposure, famine and disease, especially cholera and typhoid, which are so easily spread when water supply and sewerage systems break down. Developing countries are particularly at risk, due to the low average level of education, poor reserves and inadequate medical services. Frequently, earthquakes occur in mountain zones, where communications are normally inadequate and may be completely cut by the earthquake. Help from the central administration in the form of blankets and tents, food, water and medicines may arrive too late. Even today, with the sophisticated air-transport system of the developed world available, thousands may die before a relief operation can be mounted.

7.5.3   Tsunami

It would be inappropriate to conclude a section on the effects of earthquakes without some reference to tsunami. The creation of 'tidal waves' by caldera collapse has already been mentioned in connection with the eruption of Krakatoa. Most tsunami, however, are created by earthquakes. Among the best documented in recent years is that following the 1964 Alaskan earthquake.

The earthquake struck at 5.36 pm on 27 March 1964. The epicentre lay about 150 km east of Anchorage near the shore of Prince William Sound. The intensity was probably around 8.4 on the Richter scale, but the earthquake was particularly effective because the main movements seem to have lasted for around 4 minutes (the duration of the 1906 San Francisco earthquake was about 1 minute). Considerable geological alterations occurred all over Alaska, with changes to relative level around the coast, subsidence and the initiation of landslides. Damage to property in towns (e.g. Anchorage) was extensive, but the low density of population resulted in a low death toll (114).

Even so, most of these deaths and much of the damage to property were caused not by the earthquake but by the subsequent tsunami. The moving force for the tsunami seems to have been the sudden displacement of water in the Gulf of Alaska. Great sea waves spread out from the gulf, travelling at high speed. Waves over 10 m high overwhelmed the coast of Kodiak Island, but most of the force seems to have been directed to the southeast as a series of at least five separate waves advanced down the western coast of Canada and the United States. Coastal buildings were destroyed, and boats capsized and sunk. At least one wave still had a height of 7 m by the time it reached California, where twelve people were drowned at Crescent City, despite a tsunami warning. The smaller effects of the waves stretched much further than that; waves of 2 m reached Hawaii, and 1 m waves were recorded in Antarctica.

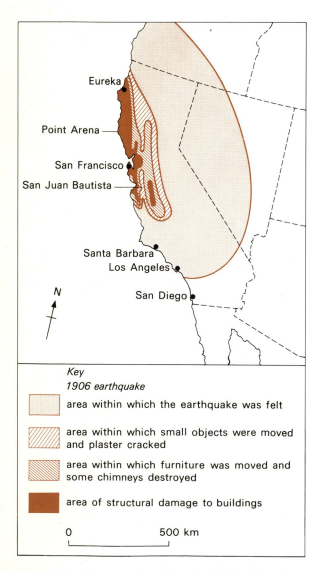

Figure 7.10   The area affected by the 1906 San Francisco earthquake (from Anderson 1971).

Appendix A

Rocks and Minerals

The rocks that may be found at the Earth's surface appear in a bewildering variety. It is normal to distinguish between: **igneous** rocks, formed from the cooling of molten material originating deep within the Earth; **sedimentary** rocks, formed by the compression and cementing of layers of sediment deposited at the surface; and **metamorphic** rocks, created by the alteration of existing igneous or sedimentary rocks under changing conditions of temperature and pressure (Section 6.2). Each of the three main groups may be further subdivided, but ultimately this great variety is produced by mixing a relatively small number of **minerals** in varying proportions. A mineral is therefore a more fundamental unit of Earth material than a rock. A mineral has a fixed, or limited, chemical composition and definable physical properties.

A.1 Igneous rock minerals

Although igneous rocks may include a wide range of mineral types, the overwhelming bulk of any igneous rock consists of minerals made by the combination of silicon and oxygen with a metal or a base. Minerals of this type are referred to as the **silicate** minerals. In Table A.1, the main groups of silicate minerals are listed together with their approximate chemistry. By far the most important of these groups is the **feldspar** group, which accounts for a large proportion of any igneous rock. Feldspars can have a range of different chemistries; those in the calcium–sodium (Ca–Na) sequence are referred to as **plagioclase** feldspar, whereas potassium (K) feldspar is usually called **alkali** feldspar. Among the other silicates are those rich in iron and magnesium, called the **ferromagnesian** minerals (e.g. **olivine**, **pyroxene**, **amphibole**). These are normally dark-coloured minerals, as are some types of the sheet mineral group, **mica**. **Quartz** (silica) is unique among the silicate minerals in consisting solely of silicon and oxygen.

In the lower part of Table A.1 these silicate groups are arranged in order of crystallisation temperature (the so called **reaction series**). It is noticeable that different feldspars crystallise across the whole temperature range, whereas the other silicate groups are far more restricted. This crystallisation sequence is important, because igneous rocks tend to consist of silicate minerals that form at roughly the same temperature. Thus, it is common to find a rock consisting of anorthite, olivine and pyroxene, or of quartz, mica and orthoclase, but rare to find a rock within which both quartz and olivine occur.

The reaction series is also very useful because it gives a good guide to mineral resistance to weathering. Minerals that form at a low temperature (e.g. quartz, orthoclase) seem far more resistant to chemical breakdown than high temperature minerals (e.g. olivine, anorthite). The relative resistance of igneous rocks can therefore be gauged from their mineral composition.

A.2 Igneous rocks

The classification of igneous rocks (Table A.2) is based upon two characteristics of the rock: chemistry and texture.

As far as chemistry is concerned, we have already noted that different silicate minerals crystallise at different temperatures and that igneous rocks tend to consist of minerals from a limited temperature range. The table therefore uses criteria such as percentage of quartz, the type of feldspar and the type of ferromagnesian mineral to define different chemical groups. Rocks like granite and granodiorite, which are rich in quartz and are dominated by alkali feldspars, are said to be **acid**, whereas gabbro, with little quartz and mainly anorthite feldspar, is said to be **basic**.

Texture refers to the size and shape of crystals within the rock. Although there are exceptions, the size of crystals within an igneous rock normally increases with the time that the rock took to cool. This means that large masses of igneous rock, which take a long time to lose their heat, are **coarse grained** (i.e. have large crystals), whereas small masses are **fine grained**. This rate of cooling is largely determined by the type of igneous body. Table A.3 is an attempt to show the main types of igneous rock occurrences. On the one hand, igneous rocks that pour out on to the surface as lava flows are generally fine grained, whereas intrusive forms, which cool while still beneath the surface, normally have larger crystals. Within the intrusive category a distinction is usually made between minor intrusions (**hypabyssal**) and major intrusions (**plutonic**). Returning to Table A.2, igneous rocks of the same chemistry but showing different texture are given different names. Thus, basalt is the lava equivalent of intrusive gabbro, and rhyolite is the lava equivalent of granite. Hypabyssal rocks often have an intermediate texture, and the rock name is formed by adding 'micro' to the plutonic version: hence,

Table A.1   The silicate minerals.

(i) Classification and chemistry

Group	Subgroup	Example	Chemistry
feldspar	plagioclase	anorthite	$CaAl_2Si_2O_8$
		albite	$NaAlSi_3O_8$
	alkali	orthoclase	$KAlSi_3O_8$
olivine			$(Mg,Fe)_2SiO_4$
pyroxene		augite	$(Ca,Mg,Fe)SiO_3$
amphibole		hornblende	$NaCa_2(Mg,Fe)_4(Al,Fe)(Si_3AlO_{11})_2(OH)_2$
mica		biotite	$K(Mg,Fe)_3Si_3AlO_{10}(OH)_2$
silica		quartz	SiO_2

(ii) Formation and resistance to weathering

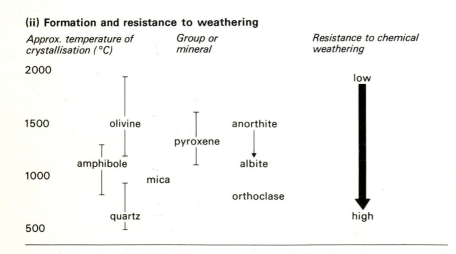

Table A.2   Classification of the igneous rocks.

Feldspar ⟶	orthoclase > plagioclase	plagioclase > orthoclase albite > anorthite	plagioclase > orthoclase albite < anorthite	no feldspar
Ferromagnesian minerals ⟶	biotite ± hornblende	hornblende + augite	olivine + augite	olivine + augite
Quarts ↓ >10% quartz	GRANITE Microgranite RHYOLITE	GRANDIORITE DACITE		
<10% quartz	SYENITE TRACHYTE	DIORITE ANDESITE	GABBRO Dolerite BASALT	PERIDOTITE –

Note on texture

Extrusive rocks (i.e. lavas) have a fine-grained texture and are shown thus: RHYOLITE.
       Intrusive rocks are generally regarded as either plutonic (i.e. major intrusions with coarse-grained texture) or hypabyssal (i.e. minor intrusions with medium-grained texture). Plutonic rocks are shown thus: GRANITE Their hypabyssal equivalents are normally recorded as Micro-. The one exception is Dolerite

microgranite and microdiorite, for example. The one exception is 'microgabbro', called dolerite.

Table A.2 does not, however, include any method for describing **pyroclastic** rocks – created by the fragmentation of lava during volcanic eruption. Pyroclastic rocks are often described by reference both to the type of lava from which they were produced as well as to the texture created by fragmentation (see Ch. 7): hence, descriptions such as andesitic ash, for example.

A.3   Sedimentary minerals

The minerals that form the sedimentary rocks fall into three broad categories.

(a)   Residual silicate minerals
Many sediments are made of fragments produced by the breakdown of igneous rocks, which are themselves made of silicate minerals. As Table A.1 suggests, different silicate minerals have different resistance to chemical breakdown. It is not very surprising, therefore, that the commonest silicate mineral found in sedimentary deposits is quartz, followed by orthoclase and mica. In contrast, ferromagnesian minerals (e.g. olivine, augite) do not usually survive prolonged weathering and transportation and are therefore rare in sediments.

(b)   Clay minerals
The chemical weathering of the silicate minerals in igneous rocks often results in the production of new minerals, the **clays**, which are really hydrous silicates, chemically quite similar to mica. Once produced, clay minerals are chemically quite stable and may therefore be transported to a sedimentary environment.

(c)   Autochthonous minerals
Residual silicates and clay minerals are transported into a sedimentary environment by some agency (e.g. running water, moving ice, wind) and therefore have their origin outside the area of deposition. (As such they are called **allochthonous** minerals.) Some minerals, however, actually form *within* the sedimentary environment and are known as **autochthonous** sediments. In most cases the material from which the mineral is made enters the environment in solution, having been produced by chemical weathering outside the environment. Once inside the sedimentary environment, chemicals in solution may be turned into sediments by biological or chemical methods. In many cases marine organisms extract calcium from the water to build their skeletal structures of calcium carbonate, which is subsequently left, after death, as the mineral **calcite** or **aragonite**. Other organisms have a silica-based

Table A.3 Classification of igneous occurrence.

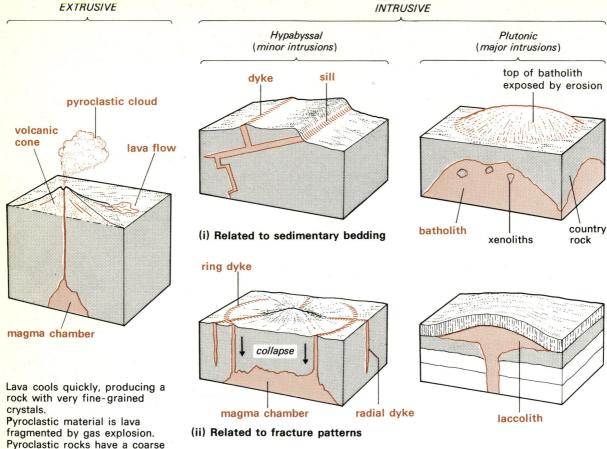

EXTRUSIVE

INTRUSIVE

Hypabyssal (minor intrusions)

Plutonic (major intrusions)

pyroclastic cloud

volcanic cone

lava flow

magma chamber

dyke   sill

(i) Related to sedimentary bedding

top of batholith exposed by erosion

batholith   xenoliths   country rock

ring dyke

collapse

magma chamber   radial dyke

(ii) Related to fracture patterns

laccolith

Lava cools quickly, producing a rock with very fine-grained crystals.
Pyroclastic material is lava fragmented by gas explosion. Pyroclastic rocks have a coarse texture, but each particle is actually a fine-grained lava fragment.

The rate of cooling being slower, hypabyssal rocks have a medium grain-size texture.

All plutonic rocks are very coarse grained due to the slow rate of cooling.

---

skeletal frame, accumulations of which ultimately form the rock **chert**. Plants generally produce carbon residues. Under certain conditions (e.g. high temperatures and limited circulation) some dissolved chemicals may be precipitated, and in deserts the complete evaporation of temporary lakes produces **evaporite** minerals. Some of the commoner autochthonous minerals are listed in Table A.4.

A.4 Sedimentary rocks

Sedimentary rocks may accumulate in a wide range of terrestrial and marine environments (see Ch. 5). Classification normally commences by distinguishing between allochthonous sediments, which consist of particles transported into the sedimentary environment from outside (**clastic** sedimentary rocks),

and autochthonous sediments, created within the environment (**non-clastic** sedimentary rocks).

(a) Clastic rocks

Clastic sedimentary rocks are primarily classified on the basis of average particle size (Table A.5) into **argillaceous**, **arenaceous** and **rudaceous** categories. Within each category further subdivision is somewhat haphazard.

Rudaceous rocks are subdivided by particle shape into angular **breccia** and rounded **conglomerate**.

In the arenaceous group (the sandstones), however, the subdivision depends upon the **sorting**, or size range, of the rock. Thus, orthoquartzites have a very uniform particle size, whereas greywackés are very mixed. In fact, this sorting usually also has a mineralogical significance. Well-sorted arenaceous sediments usually consist of residual silicate minerals that have been in a transporting system for some time, with the result that only resistant quartz is left: hence, **orthoquartzite**. An **arkose** shows less sorting and less silicate weathering and therefore includes feldspars and quartz. A **greywacké** is likely to include a wide range of minerals.

To a large extent the argillaceous rocks are composed of clay minerals, which are naturally very small in size, although some finely ground silicate material (especially quartz) may also be present. Subdivisions of this group, however, are based upon physical properties such as water content or internal structures (Table A.5). The one exception here is **marl**, which is a clay rock with an appreciable carbonate content.

(b) Non-clastic rocks

By far the most important non-clastic sedimentary rocks are the **limestone** group. Limestones are normally composed largely of the mineral calcite, but this can be fixed by either biological or chemical processes. Many limestones are broadly organic in origin, consisting of shell material in one form or another. **Chalk** is really an organic limestone, since it is made of microscopic plankton shells. A common chemical limestone is **oolite**, made of small round calcite spheres produced by wave action in shallow saturated water.

Other non-clastic rocks are economically very important (this includes the carbon-based fuels and various evaporite deposits), but they make up only a small proportion of most sedimentary sequences.

Both clastic and non-clastic sediments are unconsolidated when formed. They are turned into solid sedimentary rocks by the process known as **diagenesis**, which involves **compaction** of the layers by pressure and **cementation** of the particles by circulating mineral-rich water.

A.5 Metamorphic rocks and minerals

Metamorphism involves the alteration of rocks by heat and/or pressure. The three types of metamorphism commonly recognised (dynamic, thermal and regional) are discussed more fully in Section 6.2.3. In general, however, it is worth noting that metamorphic alteration may involve a change in the **mineralogy** of a rock or a change in its **texture**.

Most mineralogical changes are primarily due to changes in temperature. Although the processes of metamorphism do not generate temperatures which are sufficiently high to actually melt most silicate minerals, solid state chemical changes may be

Table A.4 Autochthonous sedimentary minerals.

Group	Mineral	Chemistry	Environment
carbonate	calcite	$CaCO_3$	marine or lacustrine
	aragonite	$CaCO_3$	
	dolomite	$CaMg(CO_3)_2$	
	siderite	$FeCO_3$	
silica	chert	SiO_2	
	flint	SiO_2	
oxide	haematite	Fe_2O_3	oxidising marine or lacustrine
	limonite	$HFeO_2$	
sulphide	pyrite	FeS_2	reducing marine or lacustrine
sulphate	anhydrite	$CaSO_4$	evaporite lake or lagoon
	gypsum	$CaSO_4 \cdot 2H_2O$	
	epsomite	$MgSO_4 \cdot H_2O$	
chloride	halite	$NaCl$	
	sylvite	KCl	

accelerated and thermally unstable materials such as the clay minerals may show radical alteration. Since the clays are produced by the chemical weathering of silicate minerals, it is not perhaps surprising that a rise in temperature causes the alteration of clay to related minerals such as **chlorite** or eventually back to silicates such as mica. At higher temperatures and pressures new silicate minerals, not commonly found in igneous rocks are created. These include **garnet**, **andalusite**, **kyanite** and **sillimanite**. To some extent, these minerals can be used as a guide to the temperatures and pressures attained during metamorphism (Fig. 6.20). On the whole, however, rocks which are poor in clay minerals show relatively little mineralogical change.

A quartz-rich sedimentary rock, for example, may look very similar after metamorphism although it may now be called a **meta-quartzite**. Rocks rich in calcite, such as limestone or chalk, will look different after metamorphism but the **marble** so produced is liable to be mineralogically identical.

Textural changes may also be due either to temperature or to pressure. Where heat but no pressure is applied, the metamorphic rock produced is often a rather amorphous mass. Clay-rich rocks, for example, form **hornfels** which is generally an undifferentiated mass of quartz, mica and feldspar. At the simplest level, the sudden application of stress, at a fault for example, may cause fragmentation of the rock. **Fault breccia**, consisting of

Table A.5 Classification of the sedimentary rocks.

(i) Clastic rocks

(allochthonous sediments originating outside the sedimentary environment)

Category	Size range (mm)	Subdivision	Basis of subdivision
rudaceous	>2.0	breccia	angular particles
		conglomerate	rounded particles
arenaceous	0.0625–2.0	orthoquartzite	good sorting, quartz dominant
		arkose	medium sorting, quartz and feldspar dominant
		greywacké	poor sorting, mixed mineralogy
argillaceous	0.0039–0.0625	siltstone	average size
	<0.0039	mudstone	average size, dry, solid
		clay	high water content, plastic
		shale	brittle, platy structure
		marl	mudstone with carbonate

(ii) Non-clastic rocks

(autochthonous sediments formed within the sedimentary environment)

Group	Main rock types
carbonates	limestones of either biological or chemical origin
	dolomite
ironstones	combinations of sedimentary oxides and sometimes carbonates
evaporites	anhydrite or gypsum, halite and potash sequence
siliceous	chert and flint layers, often in limestones
carbonaceous	fossil fuels (e.g. coal, oil shales)

otherwise unchanged angular fragments of the surrounding rocks, frequently lines a fault. Under special conditions the fragmentation may produce a ground rock flour. **Mylonite** is a specific type of rock flour which has subsequently recrystallised into a solid mass. Continued pressure from one direction may cause minerals to align themselves with the long axis of the crystal at right angles to the direction of pressure. Clay-rich rocks, for example, easily show **foliation** which gives **slate** its distinctive cleavage. Rocks rich in plate-like mica may develop a **schistose** texture in which the minerals occur in sub-parallel layers. The surface of each layer is typically made highly reflective by the alignment of the mica (this is a characteristic of the rock called **schist**). An extreme occurs where sustained pressure causes different minerals within the rock to separate into distinctive bands. **Banding** is the distinctive feature of **gneiss**.

Ultimately, the nature of any metamorphic rock will depend both upon the composition of the original rock as well as upon the changes of temperature and/or pressure to which it has subsequently been subjected. It probably goes without saying that no simple classification of metamorphic rocks means a great deal.

The Geological Time Scale

Traditionally, geologists have used words rather than numbers to describe various time periods in the past (Table B.1). The reason for this slightly confusing state of affairs is that, until quite recently, the main method of determining rock age was the fossil content of sedimentary strata. This method, of course, only allows the *relative* dating of different strata; it is possible to determine that X is younger than Y but not how old, in years, either may be. The result was that most attention was paid to the rocks that are fairly rich in fossils. Consequently, the last 600 million years were divided into three **eras** (Palaeozoic, Mesozoic and Cainozoic), each of which was subdivided into **periods** named variously after a typical location (e.g. Devonian, Permian, Jurassic), after a dominant rock type (the presence of coal resulted in the name Carboniferous) or after something completely irrelevant (Ordovician and Silurian were named after ancient Welsh tribes). All the vast stretch of time before the Cambrian was simply grouped together as the Precambrian, because little fossil evidence was available to produce a relative time scale.

The advent of the nuclear age has completely altered this situation. It is now known that a number of naturally occurring elements appear in unstable isotope form, which breaks down by nuclear fission into lighter elements. Among the most common are: two isotopes of uranium ^{235}U and ^{238}U, which break down to give lead (^{207}Pb and ^{206}Pb respectively); potassium (^{40}K), which breaks down to argon (^{40}Ar); and rubidium (^{87}Rb), which breaks down into strontium (^{87}Sr). In each case laboratory analysis has established the rate at which the breakdown occurs – a figure often given as the **half-life** of the isotope (i.e. the time required for 50 per cent of the original atoms to break down into the new forms). By measuring the proportion of **parent** and **daughter** nuclides in a mineral, it is possible to establish the age of the rock from which it comes in years (i.e. on an *absolute* time scale). The figures in Table B.1 were produced by these radiometric-dating methods.

Table B.1 The geological time scale (figures in millions of years ago).

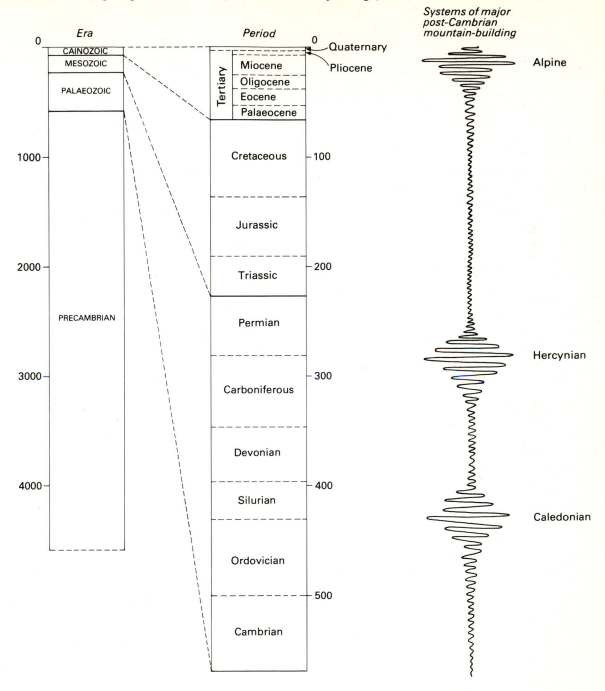

Further Reading

There are virtually no books available for Advanced Level Geography students that cover the entire field of plate tectonics and associated surface features. The best are:

S. P. Clark 1971. *Structures of the Earth*. Englewood Cliffs, NJ: Prentice-Hall.

D. C. Heather 1979. *Plate tectonics*. London: Edward Arnold.

Although not dealing with the subject from a unified plate-tectonics standpoint, much of the surface and subsurface evidence is presented in:

A. Holmes 1977. *Principles of physical geology*. London: Nelson.

A good deal of the more recent evidence is presented in a most interesting way in the early chapters of:

R. J. Rice 1977. *Fundamentals of geomorphology*. London: Longman.

In terms of detailed examples of some of the topics, presented in a very graphic way, the reader would be hard put to find anything better than:

J. T. Wilson (ed.) 1976. *Continents adrift and continents aground*. San Francisco: W. H. Freeman. Readings from *Scientific American*.

For particular areas of interest, the following are highly recommended as being written in an accessible but fascinating way:

P. Francis 1976. *Volcanoes*. Harmondsworth: Penguin.

D. H. and M. H. Tarling 1971. *Continental drift*. London: Bell.

Index